Acid-Base and Electrolyte Balance

By
Gösta Rooth, M.D. Professor
of Perinatal Medicine
Perinatal Research Unit
University of Uppsala, Sweden

A Studentlitteratur Publication
Distributed by
Year Book Medical Publishers, Inc.
35 E. Wacker Dr. – Chicago, Ill

Copyrighted © 1974 by Gösta Rooth.
All rights reserved. No part of this publication may be reproduced, stored in a retrieval system, or transmitted, in any form or by any means, electronic, mechanical, photocopying, recording, or otherwise, without prior written permission from the publisher.
Printed in Sweden
Studentlitteratur
Lund 1976

Distributed in Continental North, South and Central America, Hawaii,
Puerto Rico, and the Philippines by
Year Book Medical Publishers, Inc.
(ISBN: 0-8151-7427-6)
by arrangement with Studentlitteratur
Library of Congress Catalog Card Number: 74-29349

Table of content

Introduction 5

Chapter 1
Terminology 7

Chapter 2
Introductory aspects of acid-base balance 9

Chapter 3
On water balance 15

Chapter 4
On acids 20

Chapter 5
Current methods for determination of acid-base balance 22
    The equilibration technique of Astrup 22
    The P_{CO_2} electrode 27
    Automatic methods 28
    Technical notes 28

Chapter 6
The regulation of pH and causes of acidosis 30

Chapter 7
The relation between the water-electrolyte- and acid-base balance 34
    Intracellular electrolytes 40

Chapter 8
Buffering and Base excess 42

Chapter 9
Calculation of acid-base parameters 46
    The Siggaard-Andersen curve nomogram 46
    The Siggaard-Andersen alignment nomogram 48
    Estimation of acid-base status from inadequate measurements 55
    The slide rule of Severinghaus 56
    BE/P_{CO_2} diagram 57
    P_{CO_2}/BE diagram 58
    Formulas for calculations 58

    Corrections for unsaturated haemoglobin   59
    The use of capillary or venous blood   59

Chapter 10
Respiratory insufficiency   61

    Renal compensation for respiratory acidosis   63

Chapter 11
Metabolic compensation for respiratory acidosis   64

    Clinical example of metabolic compensation for respiratory acidosis   70
    Clinical example of metabolic compensation for respiratory acidosis + iatrogenic metabolic alkalosis due to diuretics   71

Chapter 12
Metabolic compensation for respiratory alkalosis   73

Chapter 13
Respiratory compensation for metabolic acidosis and alkalosis   75

Chapter 14
Diabetic acidosis   77

    On acidosis   77
    On potassium   81
    On water   83
    Hyperosmolar nonketotic diabetic coma   86
    Lactic acidosis   87

Chapter 15
Renal insufficiency   90

    Clinical conditions in which the acid-base regulation of the kidneys is reduced   96

Chapter 16
Acid-base disturbances due to electrolyte changes   100

    Clinical example of metabolic acidosis   102
    Clinical example of metabolic alkalosis   104

Chapter 17
Diuretics   107

    The influence of diuretics on the electrolyte and acid-base balance   107
    The need for supplementary potassium during therapy with diuretics   108

Chapter 18
Salicylate intoxication   110

References   112

Introduction

This book is based on my two earlier books on the same subject: Acid-Base and Electrolyte Balance Vol 1 of which appeared in 1966 and Vol 2 in 1970. A comprehensive Swedish edition was requested for use by Swedish students and published in 1974. The present volume is the English version of the rewritten work, with the addition of chapter 2. The repeated reprinting of the earlier books shows the interest in this short presentation in spite of the number of other books available on the subject.

Over the years I have taught Acid-Base and Electrolytes to a large number of students in different medical schools. In the present edition I have taken into account those problems which the students have generally had difficulties with, or special interest in. I have also eliminated some steps in the evolution of the subject matter and added new ones, the most important being Base Excess$_{ECF}$.

Again I have tried to teach *principles*. When examples are given they are used to illustrate principles; I have not intended to cover the subject clinically as, knowing the variety of Nature, this would seem futile. However, the number of principles to be learned is limited and if the reader masters these, he or she will be able to deal with most cases.

Some changes have been made in terminology in order to comply with current usage. For example "milliequivalents" is exchanged for "millimols" whenever possible. Pco_2 and Po_2 are expressed in mm Hg, but pressure will be given in Pascal in the future (now and then both terms are used in order to acquaint the reader gradually with the coming usage). Although pH is consistently employed, H^+ is also given, for the same reason. The hydrogen ion concentration should be written $[H^+]$ but I follow the clinical usage in writing only H^+.

Doctor Lars Wibell of the Department of Medicine, University of Uppsala, helped with the new chapter on kidney diseases, for which I am most grateful.

Perinatal Research Unit
University Hospital
Uppsala, January 1974
Gösta Rooth

CHAPTER 1

Terminology

Acid-Base Balance or *Acid-Base Status*. A quantitative description of pH and the factors which influence pH.

pH The negative logarithm of the hydrogen ion concentration.

H^+ Hydrogen ion concentration is given in nanomol/litre (10^{-9}).

Pco_2 = The partial pressure of CO_2 given in mm Hg. It is directly related to the amount of dissolved CO_2. In future, pressures will be given in Pascal and Pco_2 40 mm Hg = 5.3 kPa (kilo Pascal). The factor for recalculating mm Hg to kPa is 4/30 or 0.133.

Buffer line = That line in a log Pco_2/ pH diagram which describes the relationship between log Pco_2 and pH in one given blood sample.

Respiratory component of the acid-base balance = Pco_2.

Respiratory acidosis. Primary alveolar hypoventilation with a $Pco_2 > 40$ mm Hg (> 5.3 kPa) which leads to a low pH (< 7.40 or $H^+ > 40$ nanomol/litre).

Respiratory alkalosis. Primary alveolar hyperventilation with a $Pco_2 < 40$ mm Hg (< 5.3 kPa) leading to a high pH (> 7.40 or $H^+ < 40$ namomol/litre).

Metabolic component of acid-base balance = non respiratory component. All factors except Pco_2 which influence pH.

Base excess (BE). The best parameter for the metabolic component. If Base excess is negative, i.e. if there is a metabolic acidosis, it is best to use the term Base deficit (BD). BE and BD are given i mmol/litre. As the commercially available nomograms use only Base excess, negative BE values will now and then be given in the text. For example: Base excess was read as —18; thus Base deficit was 18 mmol/litre.

Base excess is defined as that amount of base which is needed to restore pH to normal at Pco_2 40, a temperature of 37°C, and at the actual oxygen saturation of the blood.

Note. If the equilibration technique of Astrup is used, all values will be given at saturated haemoglobin. As this leads to an increase in BE if the original blood sample contained unsaturated haemoglobin, it should always be stated if the BE values are obtained after equilibration.

Metabolic acidosis. Primary process with Base deficit which leads to lowering of pH.

Metabolic alkalosis. Primary process with Base excess which leads to elevation of pH.

Buffer base (BB) also in mmol/litre, another parameter for the metabolic component of the acid-base balance needed mainly when comparing the acid-base and the electrolyte balance.

If there is no metabolic acid-base disturbance, BB_p is 42 mmol/litre (provided the plasma protein concentration is 72 g/litre).

There is a simple relationship between BE_p and BB_p:
$$BB_p = BE_p + 42$$
This shows that changes in BB_p are equal to changes in BE_p or:
$$\Delta BB_p = \Delta BE_p$$
Increased BB values are seen in metabolic alkalosis and *decreased BB* values occur in metabolic acidosis.

Respiratory compensation for a primary metabolic acid-base disturbance should be described as follows:

as compensation for the metabolic acidosis, Pco_2 was reduced to xx mm Hg or, as compensation for the metabolic alkalosis, Pco_2 was increased to xx mm Hg.

Metabolic compensation for primary respiratory acid-base changes should read as follows:

as compensation for the respiratory acidosis, BE_{ECF} was increased xx mmol/litre or,

as compensation for the respiratory alkalosis, BE_{ECF} was reduced to xx mmol/litre (alternative: BD_{ECF} was increased to xx mmol/litre).

CHAPTER 2

Introductory aspects of acid-base balance

Acid-base balance is reputedly a difficult subject to learn. To this I can add that it is also difficult to teach because nowhere in the curriculum is there adequate time for both the mastery of basic knowledge and the study of clinical examples. Anyone who wants to master the subject is therefore forced to read some textbook. The fact that so little time is available also for reading has centered the teaching on techniques of measurement, calculation and interpretation and has left very little space for a proper understanding of the underlying physico-chemical and physiological reactions.

The tremendous advantages in the area of acid-base balance which have made pH and P_{CO_2} measurements one of the most frequent and important analyses in clinical medicine today are of course due to the new techniques and methods of interpreting the results. In scientific papers, basic knowledge of the subject is taken for granted; therefore a textbook on acid-base, electrolyte and water balance, with a description of the basic factors behind the apparent intricacies of acid-base problems, is needed.

The physiological acid-base problem lies in the fact that a large amount of hydrogen ions are metabolically produced in the cells but the hydrogen ion concentration must, in spite of this, be kept low and within strict limits otherwise the cells themselves will die. Nature solves the problem by: (1) **keeping the H^+ elimination equal to the metabolic production**, (2) **buffering H^+ so** that the free hydrogen ion concentration is kept within the physiologically acceptable margins.

The main hydrogen ion production by metabolism which we are concerned with here is that resulting from the metabolism of carbohydrates, fat, and protein resulting in CO_2. Glucose, for instance, is converted to CO_2 according to the following overall reaction:

$$C_6H_{12}O_6 + 6O_2 \rightarrow 6CO_2 + 6H_2O \tag{1}$$

CO_2 then reacts with water:

$$CO_2 + H_2O \Leftrightarrow H_2CO_3 \Leftrightarrow H^+ + HCO_3^- \tag{2}$$

The H^+ in reaction (2) is the metabolically produced H^+ which the living organism must eliminate, again in the form of CO_2. But balancing the vast amount of hydrogen ions in such a way that the body is unharmed — that constitutes the basic acid-base problem. It will be seen that the end product of the metabolic reaction was CO_2 and again the lungs eliminate CO_2, but in the process of transportation of CO_2 from the cells to the alveolar air, hydrogen ions are produced and may become harmful. The transport system accepts CO_2 in the tissues and ejects it into the alveolar air, but while in transport the CO_2 has reacted according to (2) and the resulting hydrogen ions must be kept within the acceptable limits of concentration.

In Fig 1 it is shown that the metabolically produced CO_2 reacts with H_2O in the cells to form H^+ and HCO_3^- and the same occurs in the interstitial fluid, in the plasma, and in the red cells. However, the transport of CO_2 across the cellular or vascular membranes is mainly in the form of CO_2 and not in the hydrated and dissociated form. It follows that in the tissues where CO_2 is produced the reaction (2) takes place two times; the same holds true in the capillaries of the lungs.

Looking again at the reaction $CO_2 + H_2O \Leftrightarrow H_2CO_3 \Leftrightarrow H^+ + HCO_3^-$ it will be seen that an equal amount of H^+ and HCO_3^- is produced, but it must be remembered that in the body fluids the concentration of H^+ is very much lower than that of HCO_3^- because of the buffer systems.

Buffering. A combination of a weak acid and a strong base or of a strong acid and a weak base is called a buffer solution because, given appropriate concentrations, the addition of hydrogen or OH^- ions has little effect on the pH of the solution. By mixing the acid and the base in different proportions and measuring the resulting pH, a titration curve is obtained, as illustrated in Fig 2. Here 1 ml of 0.01 N $NaHCO_3$ (a weak base) was added repeatedly

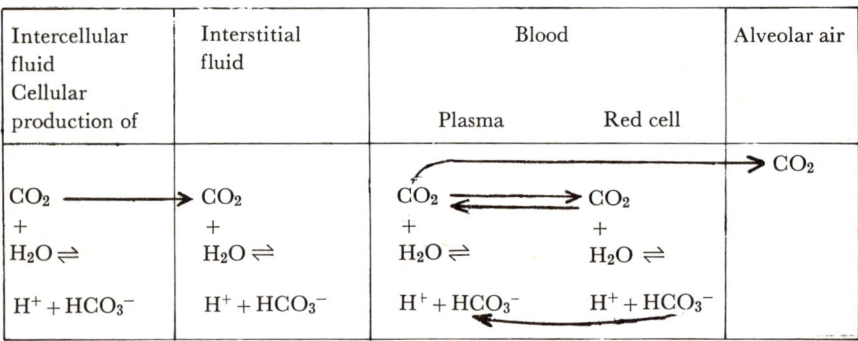

Fig. 1. CO_2 is produced in the cells and passes as such between the different membranes, but in each compartment CO_2 reacts with H_2O according to (2). The intermediary formed H_2CO_3 is deleted here.

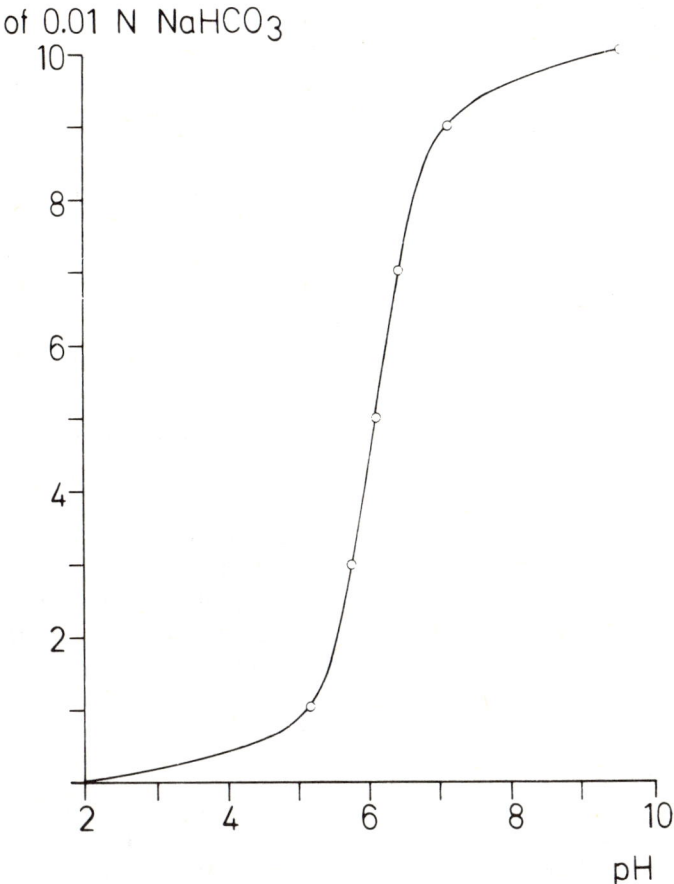

Fig. 2. Buffer curve or titration curve.
The addition of 0.01 N NaHCO₃ to 10 ml of 0.01 N HCl.

to 10 ml of 0.01 N HCl (a strong acid) and the pH was measured after each ml. The initial pH was 2.0. The addition of the first ml of NaHCO$_3$ increased the pH to 5.1. However, the addition of the same amount of base when the pH was 6 changed the pH by only about 0.1 pH unit. Thus at this range the pH was little affected by the addition of base (and of acid). We have here a buffer. You will note that this so-called titration curve has its steepest point at about pH 6 and that the steep part covers 6±1 pH unit, i.e. outside 5 and 7 the buffering action is small.

Either experimentally, as illustrated in Fig 2, or by means of the law of mass action, it may be shown that the maximal buffering action is at the pK of the solution, where K is the dissociation constant and pK is the negative logarithm of the dissociation constant.

The buffering effect is usually quantitatively expressed as $\Delta pH/\Delta H^+$, i.e. the pH change noted for each added unit of hydrogen ions.

A weak acid dissociates into H^+ and a base; the latter is called the conjugate base of the weak acid. H_2CO_3 dissociates into H^+ and HCO_3^-. HCO_3^- is therefore the conjugate base of H_2CO_3. If we now have 20 ml of a 0.01 N $NaHCO_3$ solution and add 1 ml of 0.01 N H_2CO_3 solution, the pH is found to be 7.4, the same value as in normal plasma. We will return to this shortly, but first let us observe the effect of further addition of H_2CO_3. The curve in Fig 3 illustrates the pH change until we have added 20 ml of H_2CO_3. At this point the amounts of H_2CO_3 and $NaHCO_3$ are equal and the pH is 6.1.* You will also see that when pH increases towards 7.40, the buffering capacity of the bicarbonate system actually decreases. Put in another way, the efficacy of the bicarbonate buffer system increases when the normal pH decreases.

In the blood the main buffer is the bicarbonate system, as depicted in Fig 3, and the normal pH of whole blood and plasma is 7.40. It follows that the relation between the HCO_3^- and the H_2CO_3 concentrations must be 20:1, as at the point of the curve where 1 ml of H_2CO_3 was added to 20 ml of $NaHCO_3$. The actual concentration of H_2CO_3 in plasma is 0.0012 N or 1.2 mmol/litre and that of HCO_3^- is 24 mmol/litre.

To return to the CO_2 production we know that at a normal metabolic rate in an adult, about 10 mmol CO_2 is produced per minute. In contrast to HCl, which is totally dissociated in H^+ and Cl^-, H_2CO_3 is only partially dissociated with a pK of about 4 and the amount of free H^+ is therefore less than the amount of CO_2 which enters the reaction. If all the CO_2 had dissociated into H^+ and HCO_3^- and the 10 mmol produced per minute had been evenly distributed over the total body water of approximately 50 litres, its H^+ concentration would become $1/5$ mmol/l $= 0.2 \cdot 10^{-3}$ mol/l. However, the actual H^+ is only $40 \cdot 10^{-9}$ mol/l (which is just another way of writing pH = 7.40).

The reaction $CO_2 + H_2O \Leftrightarrow H_2CO_3 \Leftrightarrow H^+ + HCO_3^-$ is therefore to some extent misleading in that it does not show how H^+ is bound to the buffer systems and how HCO_3^- is balanced against Na^+ (and K^+).

It is actually a practical impossibility to produce a correct illustration of the quantitative relationships, as the concentration of H^+ is only about 1/1,000,000 that of the buffering HCO_3^- and $Prot^-$. Fig 4 illustrates the main buffer systems in plasma. The concentration of HCO_3^- is 24 mmol/l and that of

---

* Working with pure solutions the pH would actually become 6.3, but the pK' for the bicarbonate system of the blood is used here also.

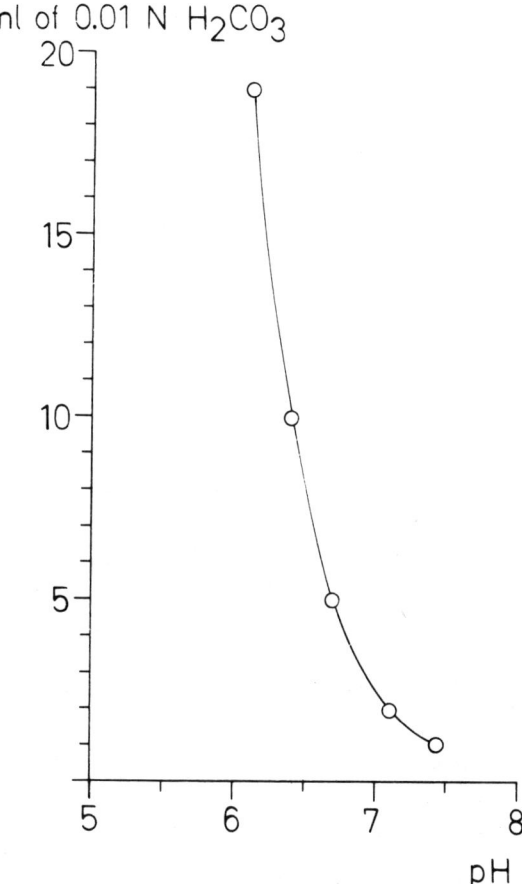

Fig. 3. How H_2CO_3 is buffered by a $NaHCO_3$ solution.
The addition of 1–20 ml of 0.01 N H_2CO_3 to 20 ml of 0.01 N $NaHCO_3$.

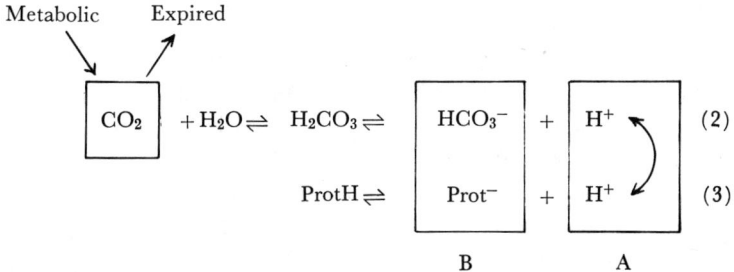

Fig. 4. The bicarbonate and protein buffer systems.

Prot⁻ about 17 mmol/l, i.e. the total buffering capacity is 41 mmol/l or 41 10^{-3} mol/l, but that of H^+, as stated above, is only 40 10^{-9} mol/l.

Fig 4 describes how H^+ is transported and is most useful in explaining the fundamentals of the acid-base balance. There will be cause to return to this set of equations later but it is worth-while here to study them more in detail. A in Fig 4 gives the hydrogen ion. It will be seen that if they are generated from the metabolic CO_2 production, they pass from equation (2) to equation (3) and of course the hydrogen ion concentration is the same in both. To express it differently, metabolically generated CO_2 drives equation (2) to the right and equation (3) to the left. It will also be noted that the same reaction occurs whether the metabolic production of CO_2 increases (as during muscular work) or the CO_2 elimination decreases (as during respiratory insufficiency). A steady state is maintained only as long as the expired CO_2 equals the metabolic production of CO_2.

If by contrast the hydrogen ions come from any other source than the metabolic CO_2, say from lactic acid produced in excess during hypoxia or 3-hydroxybutyric acid produced in excess in diabetic ketosis, both reactions (2) and (3) are driven to the left. The concentration in B, i.e. both HCO_3^- and Prot⁻, decreases and ProtH and CO_2 is produced. The latter is the same phenomenon seen when acid is poured into beer: excess CO_2 is liberated in ample foaming. Some of the CO_2 does not react with H_2O but is physically dissolved and corresponds to Pco_2.

It will also be seen from Fig 4 that if the bicarbonate or the protein concentration is lower than normal, the buffer capacity of the plasma will be reduced and addition of hydrogen ions will then raise the hydrogen ion concentration more than is usual. How the bicarbonate concentration is regulated by the kidneys will be described later.

The acid-base balance deals with how these different factors H^+, HCO_3^- + Prot⁻, and CO_2 are balanced against each other. The famous Henderson-Hasselbalch equation is nothing but a logarithmic transformation of equation (2) applying the law of mass action and taking into account the dissociation constant of H_2CO_3 and the solubility coefficent of CO_2. But it will be realized from Fig 4 that a system which comprises both equations (2) and (3) will better describe the acid-base balance than the bicarbonate system (equation 2) alone. It will be the aim of the following presentation to describe a system which takes into account both the bicarbonate and the protein buffers and which at the same time is based on clinical measurements.

CHAPTER 3

On water balance

Man gets water

1 by drink
2 by food
3 by metabolism

and loses water

1 by urine
2 by respiration
3 by perspiration
4 by stools

It goes without saying that the individual factors vary greatly from person to person and from day to day. However, it is useful to know the order of magnitude of the different factors under normal conditions.

A schematic drawing (Fig 5) illustrates that drink intake more or less equals urine output and that water from food approximately equals water lost by respiration and perspiration. The metabolic water production is about 300 ml/24 hours and the water loss in the stools about 100 ml/24 hours.

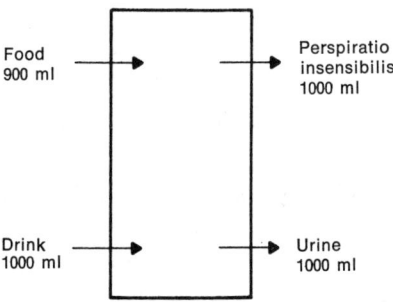

Fig. 5. Schematic drawing of the water intake via food and (drink) and water loss via the lungs, the skin, and via urine.

In caring for a patient who cannot drink properly, it is not enough to restore the fluid intake to the usual volume of drink. It must be remembered that such a patient also cannot eat properly and that the amount of water normally contained in the food intake must also be compensated for. In addition, one must remember the extra water loss due to hyperventilation caused by increased body temperature and metabolism and increased perspiration during fever.

Many water balance sheets in use give a somewhat misleading sense of accuracy and reliability. For instance, if the patient does not eat on a particular day, he may get too little fluid, or if he does eat, perhaps the water contained in the food is not entered on the balance sheet.

The following figures give some indication of the relationship between the caloric intake and the water balance:

Food contains about 400 ml of water/1000 calories.
Some 120 ml of water are produced metabolically/1000 calories.
About 40 ml of water are lost in the stools/1000 calories ingested.

Over and above the water balance sheet, two details should always be specially attended to;

1. Check that the urine volume is >1 liter/24 hours.

2. Weigh the patient. Provided the patient is adequately hydrated from the beginning, he should more or less maintain his weight, but may lose 100—300 g/24 hours. In the case of kidney disease, weighing is particularly important and is definitely the main guide to water balance. (See also chapter 15.)

Perspiration insensibilis, which actually is the sum of the water loss from both breathing and sweating is about 15 ml/kg/24 hours in adults and increases 13 percent with each centigrade of fever.

In newborn infants perspiration insensibilis is at least twice as great as in adults, i.e. 30 ml/kg/24 hours.

Muscular work, fever, salicylate intoxication, thyreotoxicosis, or other factors which increase the metabolism also increase the water loss. In individual cases it is hardly feasible to measure water lost through perspiration, but water lost via respiration may be calculated from the alveolar ventilation and the body temperature.

The importance of fluid ingestion during muscular work was illustrated in a study by Staff & Nilsson (41) who showed that work at the level of 70 % of maximal capacity could be better maintained if the subject drank 225 ml of water every 20 minutes, i.e. about 0,7 liter per hour. It is also a well known fact from mountain climbing expeditions that the participants must drink much. Because of the combination of intense hyperventilation and strenuous work, the mountaineers lose so much water that they must drink several litres per day.

Some 60 % of the body weight consists of water; this is called the total body water.

Total body water is distributed into the following compartments:

1. Extracellular fluid compartment consisting of
   a plasma
   b interstitial fluid compartment
2. Intracellular fluid compartment

In adults, plasma constitutes about 1/3 of the extracellular fluid compartment and represents 5 % of body weight,
Interstitial water is 12 % of body weight,
extracellular fluid water is 17 % of body weight and
intracellular water 43 % of body weight.

These figure vary with age, sex, and amount of body fat.
In newborn infants total body water is 75 % of body weight.
In infants below 1 year of age total body water is 65 % of body weight.
In children 1—10 years of age total body water is 60 % of body weight.
In men below 40 years of age total body water is 60 % of body weight.
In men 40—60 years of age total body water decreases to 50 % of body weight.
In women under 40 years of age total body water is 50 % of body weight.
In women 40—60 years of age total body water decreases to 45 % of body weight.

The different fluid compartments are separated by semipermeable membranes as illustrated schematically in Fig. 6. Without these membranes there would be only one single body water compartment and the different compositions of the three main types of compartments could not be maintained.

The main difference between plasma and interstitial fluid is that plasma contains about 70 g protein/liter. This protein cannot pass the capillary endothelium into the interstitial fluid. As the latter lacks the 17 mmol/litre protein anions present in the plasma, the other anions must make up for these 17 mmol/litre by having a concentration 1.05 times greater and the cations of the interstitial fluid are 1.05 times less than in the plasma.

The composition of the intracellular fluid is markedly different. The cell membrane is more complicated than the vascular endothelium and, among other functions, it actively transports sodium out of the cell. Consequently the intracellular sodium concentration is relatively low and that of potassium high. The intracellular chloride concentration is very low and its dominant anions are phosphate and protein. (See page 41.) A common feature of the different types of body fluid compartments is that they have the same osmolarity, i.e. 285 mOsm/litre.

Fluids with osmolarity equal to that of the body fluids are called isotonic.

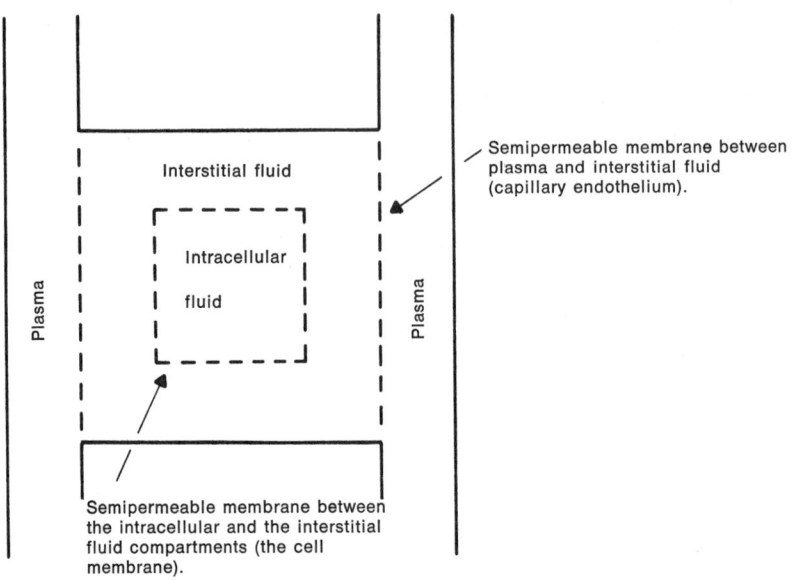

Fig. 6. Schematic drawing of the different types of body fluid compartments and the semipermeable membranes that separate them.

Fluids with osmolarity greater than that of the body fluids are called hypertonic.

Fluids with osmolarity less than that of the body fluids are called hypotonic.

Thus, even if the different fluid compartments have different concentrations of salts or other dissolved substances, the total number of dissolved particles is equal in all compartments, as the osmolarity depends upon the number of particles in solution. If the osmolarity changes in any one compartment, water moves from one fluid compartment to the others until a new osmotic equilibrium is reached.

Example: In the treatment of cardiac arrest, 100 ml of 0.6 M $NaHCO_3$ solution was administered. The patient therefore received 60 mmol Na^+ and 60 mmol HCO^-_3. This solution is dissociated to such a degree that for practical purposes we can consider that the patient received 120 mOsm. $NaHCO_3$ stays mainly in the extracellular fluid. In order to obtain the same osmolarity as the body fluids (285 mOsm/litre), these 120 mOsm must be dissolved in $120/285 = 0.4$ litre of fluid. As 100 ml were injected, it follows that another 300 ml were drawn from the intracellular fluid into the extracellular fluid compartment.

The two ions mainly responsible for the maintainance of the osmolarity of the extracellular fluid are Na^+ and Cl^-. In most cases the body osmolarity may be adequately estimated from a single sodium determination:

$$Osm/litre \sim 2\, Na^+_p$$
Normally Na^+_p is 143 mmol/litre and $2 \times 143 = 286$

In the hypothalamus there are osmoreceptors regulating the body osmolarity by releasing or retaining ADH (antidiuretic hormone).

It must always be remembered that we measure concentrations and not amounts, for instance, Na^+_p litre 143 mmol/litre. We then have one mmol of sodium in 7 ml of water. The intimate relationship between electrolytes, mainly sodium, and water must always be kept in mind.

If a patient has become dehydrated and has also lost salt, it serves no purpose to give only water as it will be immediately lost again. Only by giving water *and* salt will the fluid stay in the body. In clinical practice the patients usually lose more fluid than salt, in relation to the plasma composition or, expressed differently, the fluid lost usually corresponds to a hypotonic solution. It is therefore not necessary to give 143 mmol sodium for every litre of fluid. Instead, the isotonic salt solution should be given alternately with a glucose solution. The osmoles in the latter, of course, become metabolised to water and CO_2.

Provided there is no serious electrolyte- or acid-base disturbance which calls for a specific electrolyte substitution, any fluid given should correspond as closely as possible to that of plasma. Saline, the classic 0.9 percent NaCl solution, is isotonic, but otherwise not physiological, as it contains 154 mmol/litre Na^+ and consequently also 154 mmol/litre Cl^-. The chloride concentration is particularly high in relation to that of plasma, and it is easy to show, theoretically and clinically that each given litre of 0.9 percent NaCl will give the patient a Base deficit of 3 to 4 mmol/litre. The Ringer solution is no more physiological than saline. True, it contains 4 mmol potassium per litre, but its chloride concentration is even higher than that of saline and leads to a metabolic acidosis of the same magnitude as saline.

Newer solutions containing acetate have an ionic balance which closely resembles that of plasma and will not induce any acid-base changes. The acetate is metabolized to bicarbonate. Theoretically it would have been as well to add bicarbonate instead of the acetate to the solution, but there are Galenic difficulties with bicarbonate.

In a book of this kind it would be an advantage to present a table of the different solutions available. However, they vary from country to country, and sometimes from one hospital to another; therefore no representative list can be given.

CHAPTER 4

On acids

The present conception of the nature of acids and bases is closely related to clinical problems. It is told that when Dr Eric Warburg, who later became chairman of the Department of Medicine of the University of Copenhagen and President of that university, defended his thesis, Brönsted was the faculty opponent. He felt that the then prevailing theory of the nature of acids and bases was not adequate and subsequently formulated the conception generally accepted today that acids are hydrogen ion or proton donors and bases are hydrogen ion or proton acceptors.

Some substances are both acids and bases. The best example of this is water. H_2O can accept H^+ which gives H_3O^+, and water may give off H^+ resulting in OH^-. H_3O^+ is called the hydronium ion and represents that form in which H^+ is actually present in the body, but in clinical usage the hydration is deleted and we simply speak of H^+.

H_2O spontaneously dissociates into H_3O^+ and OH^- but in the equation
$$2H_2O \rightleftharpoons H_3O^+ + OH^-$$
the equilibrium is to the left.

The degree of dissociation, as always, depends upon the temperature. At 24°C about 1/10 000 000 mol/litre occurs in the form of H_3O^+ and an equal amount as OH^-. At body temperature, i.e. 37°C, pH=pOH at 6.8. In the treatment of patients in hypothermia one must take into account the fact that a higher pH is "normal" at a lower temperature (33). Another way to express neutrality at 24°C is to state that the hydogen ion concentration then is 100 nanomol/litre, or that pH is 7.0, when the OH^- concentration is exactly the same. In a 1 M strong acid, in which H^+ is fully dissociated, pH is 0. In a corresponding base, pOH is 0 and pH is 14; pH instruments therefore have scales between 0 and 14 pH units.

Na^+, K^+, Ca^{2+} and Mg^{2+} were earlier called bases, but according to the terminology of Brönsted they are neither acids nor bases as they neither accept nor donate protons or hydrogen ions. Many of the substances previously called bases are cations, but Buffer base represents anions (bicarbonate and protein). As these are hydrogen ion acceptors, they are bases according to Brönsted.

The maintainance of an optimal hydrogen ion concentration is biologically of greatest importance. For the proper function of many of the enzyme systems of the cells, a normal, or close to normal, pH is needed; many metabolic functions are reduced if pH is severely reduced.

H^+ is a cation and at first sight it may be surprising that H^+ is not included in the Gamblegram shown on page 36. However, the hydrogen ion concentration is so low that it cannot be represented in the picture. Na^+_p is 142 mmol/litre, but H^+_p is only 40 nanomol/litre (10^{-9}). pH may vary from 6.9 to 7.7, corresponding to 125 to 20 nanomol/litre. It follows that percentagewise the hydrogen ion changes may be considerable, although they remain small in absolute numbers.

CHAPTER 5

Current methods for determination of acid-base balance

Acid-base determinations which some 20 years ago were only occasionally performed at our hospitals are today among the most common analyses. The main reason for this is the availability of rapid methods for measurement of acid-base parameters. Within a span of two years, two new techniques were reported: the equilibration method of Astrup (20) and the blood Pco_2 electrode method of Severinghaus & Bradley (36).

Initially the Astrup method was by far the most used but today both methods seem to be equally popular. Large laboratories, which over and above Pco_2 need Po_2 values, may prefer the simultaneous measurement with Pco_2 and Po_2 electrodes, whereas other units prefer the equilibration technique as being technically the easiest.

As an introduction to acid-base studies the equilibration method is useful in that the student immediately learns to distinguish between the respiratory and the metabolic component of the acid-base balance.

Regardless of which method of analysis a laboratory employs, the calculations are usually made from the nomogram of Siggaard-Andersen. This was developed in Astrup's laboratory and is based on measurements with the equilibration technique.

The equilibration technique of Astrup

The principle behind the method is the observation that there is a rectilinear relationship between pH and log Pco_2 in any individual blood sample (Fig. 7). Thus if two points are known, the line is defined and an unknown Pco_2 may be read from an anaerobic pH measurement. For the complete analysis, a total of three pH determinations is required. Plasma or whole blood may be used.

In order to obtain two points in the log Pco_2/pH diagram, the blood is shaken (equilibrated, tonometry is performed) with two gas mixtures with different but known CO_2 concentrations and, consequently, known Pco_2 values. Besides CO_2 the gas mixture contains O_2. It follows that any primarily unsaturated haemoglobin will be saturated during tonometry.

If one of the gases contains 4.20 percent CO_2 and the barometric pressure is 760 mm Hg, its Pco_2 is $(760-47) \times 4.20/100 = 30$ mm Hg and if the second gas has a concentration of 8.50 percent CO_2, its Pco_2 is $(760-47) \times 8.50/100 = 61$ mm Hg., 47 being the partial pressure of water vapour at 37°C.

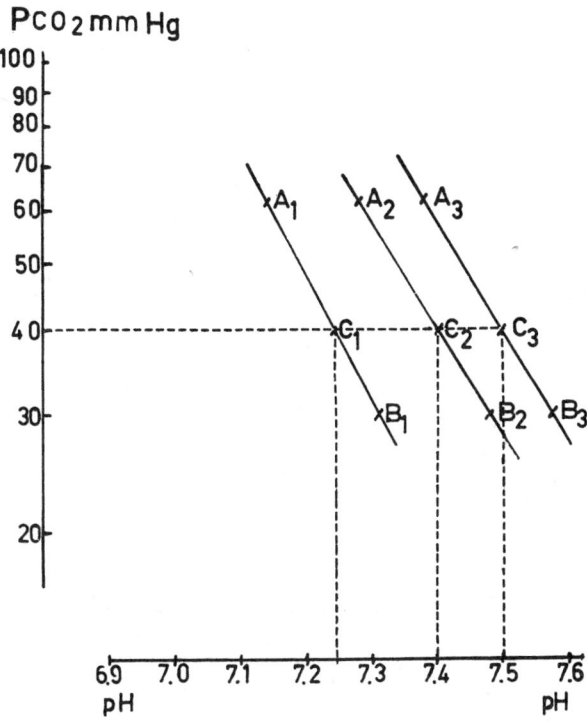

Fig. 7. The principle of the Astrup equilibration technique.
The relation between Pco_2 and pH in three different blood samples. In each sample pH is measured after equilibration with two different gases with known Pco_2 values. Thus points A and B are obtained and then connected with a straight line. This line is called buffer line and describes the relationship between Pco_2 and pH in the individual blood sample. After anaerobic pH measurement, point C is obtained. It is assumed that all three samples were taken from patients with normal Pco_2 i.e. 40 mm Hg. The actual pH i marked with the dashed vertical line. The Pco_2 values are read from the dashed horisontal line.

pH measurements on a blood sample without acid-base disturbances give pH = 7.30, using the gas with 8.50 percent CO_2, and 7.48 for the gas with 4.20 percent CO_2. Thus points A_2 and B_2 in Fig. 7 are obtained.

Before the tonometry is undertaken, pH must be measured anaerobically. In our case this pH was 7.40. Using this point and the buffer line A_2—B_2 P_{CO_2} is read as 40 mm Hg.

If acid is added to the blood, or the blood sample is taken from a patient with a certain degree of metabolic acidosis, points A_1—B_1 will be found. If again the blood is drawn from a patient with a certain amount of metabolic alkalosis, or base is added to normal blood, points A_3—B_3 are recorded. If the acidotic patient has an anaerobic pH of 7.24 and the alkalotic patient has one of 7.50, points C_1 and C_3 show that P_{CO_2} was normal (40 mm Hg) in these cases also.

It should be remembered that all respiratory changes, i.e. changes in P_{CO_2}, are represented by points along the lines A_1—B_1, A_2—B_2, and so on. The line A—B is called buffer line for P_{CO_2} because the greater the buffer capacity of the blood, the steeper the line, and for any given change in P_{CO_2}, pH will change less. Whole blood having a greater buffer capacity than plasma will be represented by a steeper buffer line.

We have already, in Fig. 7, all the necessary information for determining P_{CO_2} quantitatively, i.e. the respiratory component of the acid-base status is defined. Fig. 7 also gives information about the metabolic component, provided the respiratory component is normal, as it shows the pH due solely to metabolic changes. This information could suffice and, in analogy with Astrup's original term "Standard bicarbonate", we could speak of "Standard pH". Saling (14) calls this pH value pH_{40qu}. Unfortunately the actual pH and the "metabolic" pH_{40qu} are often confused. This is one of the reasons that it is an advantage to express the metabolic component in a different unit. Moreover, it is preferable to express the metabolic component in mmol/litre, as in the case of the other blood constituents. Futhermore, it is much easier to evaluate the therapy from a Base deficit expressed in mmol/litre than from a pH value.

In Fig. 7, two new scales may be added, one for Base excess and one for Buffer base as in Fig. 8 which otherwise is the same as before. Base excess is read where the buffer line intersects the BE scale and Buffer base where the line crosses the BB scale. Thereby the metabolic component is quantitatively determined without additional measurements as the position of the buffer line has already been determined in the P_{CO_2} determination.

Siggaard-Andersen and Engel (61) published the first of these curve nomograms, which was later revised by Siggaard-Andersen (62) on the basis

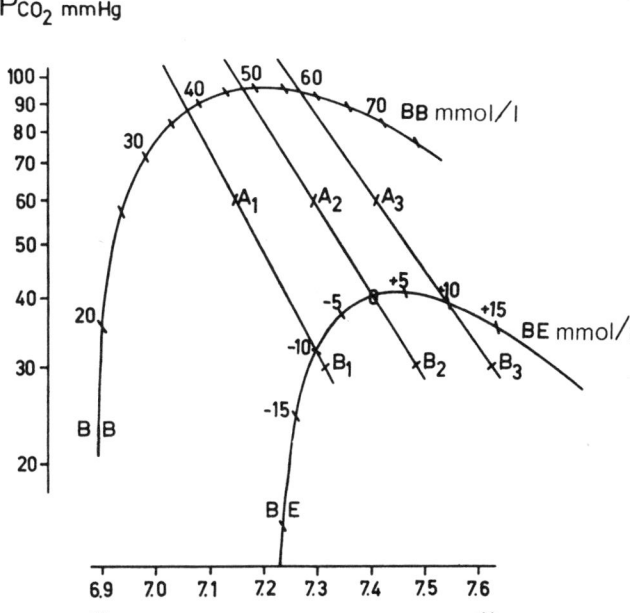

Fig. 8. Determination of the metabolic component of the acid-base balance.
This is the same log P_{CO_2}/pH diagram as in Fig. 7 with the same measurements. Two scales are added, one above for Buffer base (BB) and one below for Base excess (BE). The metabolic component is read where the buffer lines A_1–B_1 and so on intercept the BB or BE curves. The blood sample giving line A_1–B_1 has a metabolic acidosis of 10 mmol/litre and A_3–B_3 a metabolic alkalosis of 10 mmol/litre.

of new measurements. Either in the original curve form (Fig. 12) or in the linear form (Fig. 13), this is now used all over the world.

Base excess gives in mmol/litre the amount of acid (at 37°C and P_{CO_2} 40 mm Hg) which must be added to a blood sample in order to restore its pH to normal. Base deficit, consequently, expresses in mmol/litre the amount of base which must be added to a blood sample in order to restore its pH to normal (at 37°C and P_{CO_2} 40 mm Hg).

Buffer base, as defined by Singer and Hastings (64) is the sum of the anions bicarbonate and protein (compare with the Gamblegram, page 36).

As the slope of the buffer line depends upon the protein content of the blood, the slope will be a function of the haemoglobin concentration in whole blood. In the curve nomogram there is a haemoglobin scale under the BB scale (Fig. 12). This is schematically shown in Fig. 9, At Hb=0, i.e. in plasma, Buffer base is equal to $HCO_3^- + Prot^- = 24 + 17 = 41$ mmol/litre. BB increases 0.4 mmol/litre for each g/100 ml of haemoglobin, whereas Base excess is unaffected, as shown in Fig. 9, where all lines pass through BE=0.

As explained on page 44 there are several advantages in assuming that all blood samples have a buffer capacity corresponding to the Hb 5 line. Less technical errors occur, one less measurement is necessary (two instead of three), and the answer is representative for all the extracellular fluid, not only for blood. Using Hb 5, the slope of the buffer line is defined and it suffices to determine its position by one pH determination after tonometry with one gas of known Pco_2.

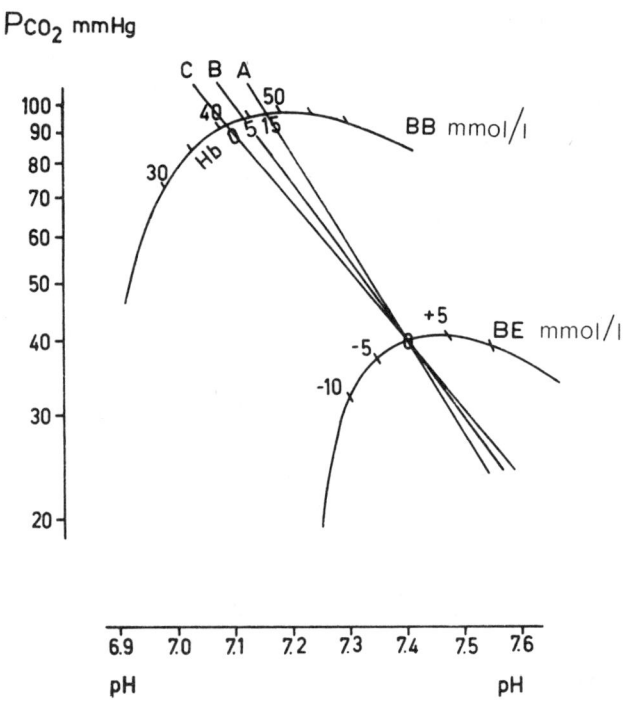

Fig. 9. How variations in haemoglobin concentration influence the buffer capacity of the blood.
This is the same diagram shown in Fig. 8. A single blood sample with no metabolic acid-base disturbance is used. Line A is obtained from whole blood with a haemoglobin concentration of 15 g/100 ml, and line C from its plasma. By mixing whole blood and plasma so that Hb 5 is obtained, line B is found. It follows that BE is not influenced by changes in buffer capacity, whereas BB changes with the protein and haemoglobin concentration.

A pure bicarbonate solution will have a more horisontal slope of its buffer line than plasma, as it lacks the buffering plasma proteins. The scales in the nomogram are chosen so that the slope of the bicarbonate buffer line is — 1.0, i.e. the bicarbonate concentration is obtained by drawing a 45° line from the actual pH to the bicarbonate scale on the Pco_2 40 level. (See Fig. 12, page 37).

Until Base excess was introduced, standard bicarbonate was the best parameter for the metabolic component of the acid-base balance. Standard bicarbonate is read where the buffer line crosses the bicarbonate scale (Fig. 12).

The Pco_2 electrode

The second current method for determining the acid-base parameters is with the Pco_2 electrode constructed by Severinghaus and Bradley (36). pH must of course still be measured with a pH electrode; therefore for a complete determination of the acid-base balance, pH equipment is always indispensible.

The Pco_2 electrode is, in principle, only a glass pH electrode but it is surrounded by a thin film of bicarbonate solution. The fluid is kept in place by a thin membrane permeable for CO_2. A gas for calibration or a blood sample is brought to one side of the membrane and Pco_2 is equilibrated across the membrane. The pH of the bicarbonate solution will be changed and may be calibrated directly in mm Hg Pco_2.

The relationship between pH, Pco_2 and the bicarbonate concentration is best seen from the Henderson-Hasselbalch equation:

$$pH = pK' + \log \frac{HCO^-_3}{S \times Pco_2}$$

Around the pH electrode the bicarbonate concentration cannot change and pH is therefore inversely related to log Pco_2, as seen from the equation.

Many textbooks still use the Henderson-Hasselbalch equation in order to explain acid-base changes. The Siggard-Andersen linear nomogram (Fig. 13) is in its pH, Pco_2 and HCO^-_3 lines nothing but a graphic representation of this equation. In clinical practice today we measure pH and Pco_2 and calculate Base excess. It then seems more appropriate to determine the relationship between these three variables than to spend time on the Henderson-Hasselbalch equation.

Automatic methods

Some producers have built Pco_2 and Po_2 electrodes together in such a fashion that a blood sample is brought into contact on one side with the Pco_2 and on the other with the Po_2 electrode. These two measurements are therefore made simultaneously. A pH meter is also added to the unit. Recently these three instruments have been built together in such a way that two, or even all three measurements are made automatically, together with the calibration. This has the advantage of increasing precision and, if many samples are measured daily, the unit may also be labour saving.

The historical background of these new methods may be of some interest. Poul Astrup was in charge of the chemical laboratory at a hospital for infectious diseases in Copenhagen when that city suffered a serious epidemic of poliomyelitis with respiratory paralysis. The mortality was high and no satisfactory respirators were available. Professor Lassen, the physician in charge, received help from the anaesthesiologist who "ventilated" the patients just as during an operation, i.e. manually by means of rubber bags. Soon the wards were filled with medical students working these bags around the clock. They succeeded in bringing a number of patients over the most difficult stage of the disease. However, in order to know when to start, how intensely to ventilate, and when to stop, it was necessary to know the Pco_2. As no convenient method was at hand, Astrup conceived his technique.

Severinghaus, who is an anaesthesiologist, was also primarily interested in Pco_2 and together with Bradley constructed the Pco_2 electrode.

Astrup and his co-workers later showed how, without further analyses, the metabolic component of the acid-base balance could be quantitatively determined.

Technical notes

A few points which have proved to be important for optimal reproducibility of the pH measurements follow:

1 If using capillary blood, there must be a free flow of blood, as any pressure will provide mainly venous blood.
2 If the analysis is not made within 15 minutes, the sample must be cooled as soon as it is obtained and may remain cooled for 4 hours.
3 Adequate mixing of the blood is always necessary. The blood samples must be well rotated immediately before measurement.
4 When not in use, the pH electrodes should be filled with a rinse and storage solution recommended by the manufacturers.

5 Before use, the pH electrodes should be washed in 0.9 % NaCl solution and then sucked dry.
6 When setting the buffers, a) fill with buffer, b) suck air, c) fill with buffer and set pH.
7 The electrode should be rinsed once each week with pepsin HCl solution.

A most useful booklet with technical notes has been prepared by Severinghaus and Bradley (37).

CHAPTER 6

The regulation of pH and causes of acidosis

In order to understand the genesis of an acidosis, or for that matter an alkalosis, attention must be given to where hydrogen ions are produced, how they are metabolized or how they are eliminated. By a systematic analysis of the fate of the hydrogen ions, virtually any acid-base disturbance can be recognized.

The hydrogen ion is of course so important in so many reactions that it would not be practical to discuss them all. It suffices to mention the most important.

The four main mechanisms causing primary acidosis are:

1. Deficient CO_2 elimination.
2. Insufficient oxygen supply to the cells leading to reduced hydrogen ion neutralization by water formation.
3. Production of abnormal acids. The most important are 3-hydroxybutyric acid and acetoacetic acid during lipid mobilization caused by diabetes or starvation.
4. Insufficient elimination of non volative acids via the kidneys.

Electrolyte disturbances may also cause metabolic acid-base derangements; for instance HCO^-_3 loss via the stools or urine may give a metabolic acidosis, as may also iatrogenic Cl^- administration. These conditions will be dealt with in Chapters 16 and 17.

The normal regulation of pH is by adequate ventilation, i.e. ad 1, by oxidation of the hydrogen ions to water, i.e. ad 2, and by elimination of hydrogen ions bound to phosphate in the urine, i.e. ad 4. These different factors will be discussed in some detail here and referred to again later.

ad 1 Deficient CO_2 elimination.

The main volume of CO_2 is produced in the metabolism of fat and carbohydrates, but CO_2 is also liberated in the breakdown of protein. For glucose the reaction is:

$$C_6H_{12}O_6 + 6O_2 \rightarrow 6\,CO_2 + 6H_2O \tag{1}$$

Some CO_2 is dissolved ($=P\text{co}_2$) but most of it reacts with water:

$$CO_2 + H_2O \rightleftharpoons H_2CO_3 \rightleftharpoons H^+ + HCO_3^- \tag{2}$$

CO_2 is produced in the cells and diffuses out into the extracellular fluid, where part of H^+ is buffered to the bicarbonate and protein, but the main buffering takes part in the red cells. Some of the bicarbonate generated in the red cells then diffuses from the cells into plasma in exchange for Cl^-.

The result is that most CO_2 is transformed into HCO_3^- and, as such, is transported to the lungs where reaction (2) goes from the right to the left. This reaction is facilitated by the enzyme carbonic anhydrase.

Fig. 10 schematically illustrates the CO_2 transport from the cells to the ambient air, going, as it does, through two types of pipe lines, i.e. the vessels and the respiratory tract, aided by one pump for each pipe line, i.e. the heart and the respiratory muscles. Normally the amount of CO_2 eliminated per time unit equals the amount of CO_2 produced metabolically. The result is that the pH remains unchanged. If the CO_2 elimination decreases, the hydrogen ion concentration increases and pH decreases. If respiration is forced, for instance because of pain, the CO_2 loss is greater per time unit than its production, the amount of dissolved CO_2 decreases and the pH increases.

The relationship between CO_2 production and ventilation is:

$$P\text{co}_2 = BP \times 22.26 \; \frac{CO_2 \text{ production}}{\text{alveolar ventilation}}$$

$P\text{co}_2$ and barometric pressure (BP) is expressed in mm Hg, CO_2 production in mol, and the alveolar ventilation in litres STP.

According to the definition, every change in $P\text{co}_2$ is called respiratory, because the most common reason for $P\text{co}_2$ changes is variation in the breathing. However, $P\text{co}_2$ may increase due to other factors as well, for instance after vigorous muscular exercise. But still we call this a respiratory acid-base change. As a result of deficient circulation, $P\text{co}_2$ may be markedly increased also.

The total CO_2 production is some 20 000 mmol/24 hours or 850 mmol/hour.

ad 2 Insufficient oxygen supply to the cells.

In the degradation of 1 mol of glucose, 2 moles H^+ are produced. If oxygen is available, water and energy are formed. Lacking oxygen, lactic acid accumulates, and for each mol of glucose, 2 moles of lactic acid are produced.

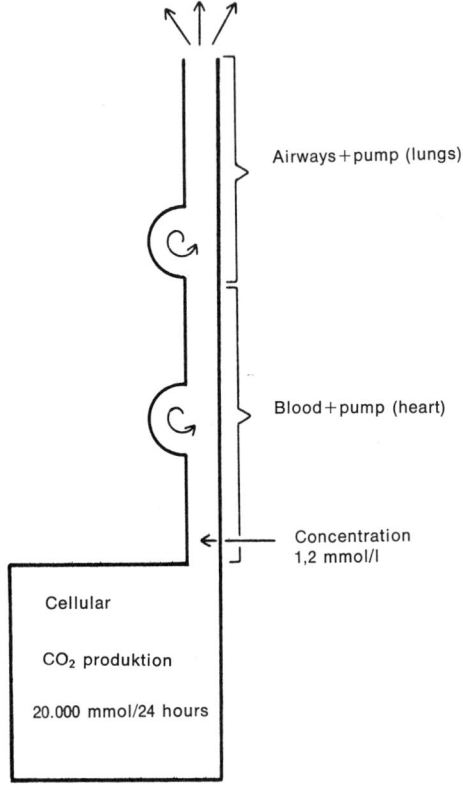

Fig. 10. Schematic drawing of CO_2 elimination.
Normally P_{CO_2} is 40 mm Hg corresponding to 1.2 mmol/litre of dissolved CO_2. As CO_2 readily diffuses through the semipermeable membranes separating the different body fluid compartments, it may often for practical purposes be assumed that P_{CO_2} is of the same order of magnitude everywhere.

The hydrogen ion is buffered, mainly to bicarbonate, CO_2 is liberated and exhaled. Lactate takes the place of the vanishing bicarbonate. Many primary factors may lead to insufficient oxygen supply to the tissues, but they all lead to lactic acid production and lactate accumulation. Such situations occur after respiratory paralysis, cardiac arrest, isometric contractions of the muscles, or when during umbilical cord compression the gas exchange of the fetus is jeopardized.

ad 3 Production of 3-hydroxybutyric acid and acetoacetic acid after rapid lipid mobilization.

The classical examples are starvation and diabetic ketosis. Normally both acids are produced and the anion residues are found in the plasma. 3-hydroxybutyrate is 0.2 to 0.4 mmol/litre and acetoacetate about 0.1 mmol/litre. When there is not enough intracellular glucose, be it because of starvation or lack of insulin, these concentrations may be increased 20—30 times. (See also page 79.)

ad 4 Insufficient elimination of non volative acids via the kidneys.

This will be dealt with separately in Chapter 15, page 90. As the kidneys normally eliminate only 100—200 mmol H^+/24 hours, a certain reduction in their capacity will only gradually affect the acid-base balance. As a rule, Base deficit increases only by a couple of mmol/24 hours and the pH changes are correspondingly small. This is in contrast to the situation in respiratory insufficiency or lack of oxygen, when pH drops rapidly within a few minutes.

CHAPTER 7

The relation between the water- electrolyte- and acid-base balance

It can not be stated too often that neither the water, nor the electrolyte-, nor the acid-base balance may be studied individually because of the strong interaction between them. Together they are govered by two physicochemical laws and the acid-base balance is also determined by a physiological law. These are:

1. The law of electro-neutrality
2. The law of iso-osmolarity
3. The attempt by the body to maintain a normal pH.

The law of electrical balance states that the sum of negative charges of the anions must be equal to the sum of positive charges of the cations. In the plasma there are 153 meq/litre of cations and consequently, according to this law, there are 153 meq/litre of anions. The law applies to anions and cations in all body compartments, as well as any other solution.

The law of iso-osmolarity states that the osmolarity is the same in the fluid systems of the body between which water is exchangeable. In other words, the osmolarity is the same in plasma, interstitial fluid, and intracellular fluid. In all the compartments it is about 285 mOsm/litre. Water moves freely between the body fluid compartments and if the number of dissolved particles (osmoles) increases in one compartment, water will move into it until a new equilibrium is reached and a new level of iso-osmolarity is attained.

Sometimes the laws of electro-neutrality and iso-osmolarity interfere with each other. This happens if there are semi-permeable membranes between two compartments, as there are in the body. Water moves freely, but the dissociated ions are more or less impeded in their transport. In this way a voltage gradient is built up between two fluid compartments. For instance, there is a difference of about 80 mV between the inside of the muscle and the extracellular fluid.

The semipermeable membranes cause the Donnan effect. One example of this is the difference in electrolyte composition between the plasma and the interstitial fluid. Plasma protein cannot pass the membrane and, as a consequence, the anion concentration of Cl^- and HCO_3^- is 1.05 times higher in the plasma, while the opposite holds true of cations, Na^+, K^+, and so on.

The third law which governs the acid-base and electrolyte balance is the physiological law saying that the body tends to restore pH towards a normal value.

As the water-, eletrolyte- and acid-base balances are governed by known laws, it is to a large extent possible to calculate the effect of therapy. However, any such calculation will necessarily be approximative, as there are factors involved of which we have insufficient knowledge, such as the exchange of Na^+ between the bones and the extracellular fluid. None-the-less, such a calculation is often worthwhile, mainly because it gives quantitative information about the rate of progress of the pathological process itself. (See also page 88.) A system for calculation is given in Rooth (12).

The best way to see the interaction between the water-, electrolyte- and acid-base data is to construct a Gamblegram (Fig. 11) which is a graphic representation of the law of electro-neutrality applied to plasma. At the same time the height of the columns gives some indication of the water balance. (Compare also with page 79.) In balancing anions against cations, the units must be milliequivalent (meq/litre). Na^+, K^+, Cl^- and HCO_3^- are monovalent and meq=mmol. Ca^{2+} and Mg^{2+} with 4.6 and 2.0 meq/litre, respectively, are bivalent and therefore 2.3 and 1.0 mmol/litre. As laboratories in the future will more and more give all their measurements in mmol/litre, this unit will be consistently used except in the present context.

The dominating cation in plasma is sodium (142 meq/litre). Potassium, calcium and magnesium, with normally 4.6 and 2.0 meq/litre respectively, add to the sum total of positive changes, i.e. 153. Then, according to the law of electro-neutrality, the sum total of anions is also 153, but it is much more complicated to measure them than the cations. Chloride is 101, bicarbonate 24, and protein 17 meq/litre. The so called residual anions follow: phosphate, sulphate, lactate and other residues of organic acids.

The sum of the bicarbonate and the protein anions is called Buffer base and amounts to $24+17=41$ meq/l. Sometimes the figure 42 is used, which is equally correct. The Buffer base in the Gamblegram is the same Buffer base as is the nomogram of Siggaard-Andersen (Fig. 12) and is therefore the connecting link between the acid-base balance and the electrolyte balance. It would be equally correct to say that the bicarbonate ion is a common factor, but, as will be shown, Buffer base is a better parameter for the metabolic component of the acid-base balance than bicarbonate. The main reason,

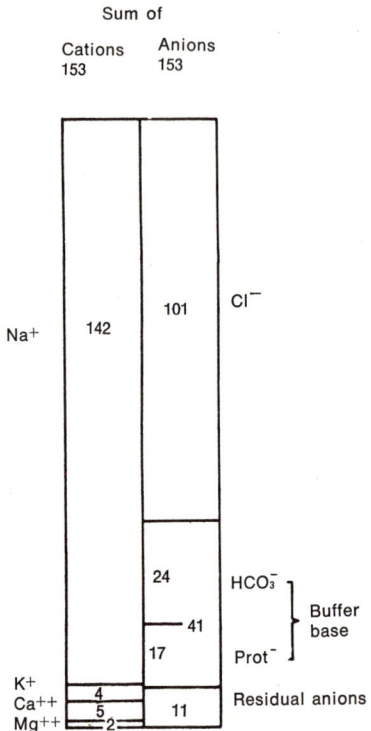

Fig. 11. Gamble diagram (ionogram).
A graphic representation of the law of electro-neutrality in plasma. The sum of the anions is equal to the sum of the cations.
Here the electrolytes cannot be given in mmol/litre as the electrical charges balance each other. Had all the ions been monovalent, this would have been without importance, but Ca^{2+} and Mg^{2+} are bivalent.

of course, is that the HCO_3^- concentration is so much influenced by the P_{CO_2} level.

Because Buffer base is represented both in the electrolyte and in the acid-base balance, it is one of the two parameters for the metabolic component of the acid-base balance which one must know.

In a situation like diabetic ketosis the patient may have a Base deficit of 20 mmol/litre. Buffer base is then $42-20=22$ mmol/litre. As the electroneutrality must still be maintained, either the residual anions in the plasma must increase correspondingly, or the cation concentration must decrease. As will be discussed in detail on page 79, in diabetic ketosis the concentration

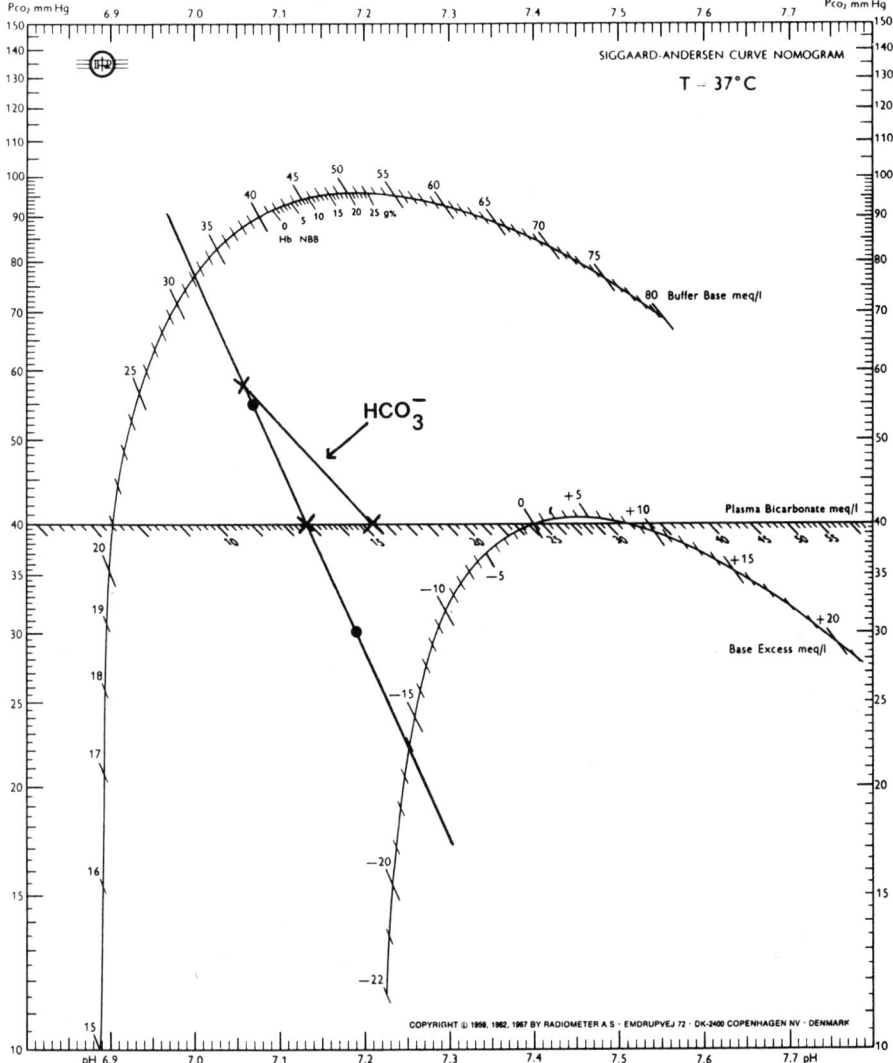

Fig. 12. The Siggaard-Andersen curve nomogram.
Acid-base data from a patient with respiratory paralysis. The circles denote pH measurements after equilibration with two different gases with known P_{CO_2} values. The cross marks the anaerobically measured actual pH value of the blood sample.
Where the buffer line cuts the P_{CO_2} 40 line there is a scale for standard bicarbonate. The actual bicarbonate concentration of the plasma is obtained by drawing a 45° line from the actual pH value to the standard bicarbonate scale.

of organic anions increases, but concomitantly there are changes in the cation and in the chloride concentrations as well.

In metabolic alkalosis, Buffer base increases and, still in order to maintain the electro-neutrality, either the chloride concentration must decrease or that of sodium increase. The other cations are not important because they cannot quantitatively change so much. The residual anions never decrease, but may be augmented in certain clinical situations (page 39). As the body also tries to maintain the osmolarity unaltered, the sodium concentration usually does not change. It follows from these arguments that the chloride level should decrease in alkalosis, which is indeed the case.

There is never a metabolic alkalosis without decreased Cl_p^-.

Schematically we may say that in metabolic alkalosis, BB_p^- increases at the expense of Cl_p^-. More to the point is the fact that the body has lost its H^+ together with Cl^-, for instance in vomiting or by the elimination of NH_4Cl in the urine. H^+ is taken from the buffer systems and its variations do not show up directly in the plasma composition as do the chloride changes. When a patient loses chloride and still has the same plasma cation concentration, another anion must substitute for it and bicarbonate is always available from the metabolism.

It is worthwhile to remember that the concentration of bicarbonate is never a direct function of its production or elimination, but depends upon the surplus of cations over anions. The sodium, potassium, chloride, and sulphate concentration cannot change rapidly and are therefore sometimes referred to as "fixed ions", whereas the HCO_3^- level rapidly adapts itself, depending upon the respiration, the buffer capacity, and the availability of cations.

If the bicarbonate concentration decreases actively, as during hyperventilation, the lactate level will increase correspondingly. Lactate is therefore less "fixed" than the other ions.

In order to become thoroughly acquainted with acid-base and electrolyte balance, it is useful to plot electrolyte data in the form of Gamblegrams. Once this is done in a number of clinical cases, the student will readily grasp the significance of the different factors. Even if all the ions are not measured, they may be given approximative values. There will be cause repeatedly to return to the Gamble diagram.

Physicians sometimes insist on complete analysis of both acid-base and electrolyte balance. Often this is superfluous. An elementary knowledge of the Gamblegram will save many unnecessary analyses.

As potassium, calcium, and magnesium together constitute about 11 mmol/litre and approximately equal the normal residual anion concentration (10 mmol/litre) the Gamblegram may often be simplified according to Fig. 25, page 80. It shows that $Na_p^+ \sim Cl_p^- + BB_p^-$, or if we solve this with regard to BB_p^-:

$$BB_p^- \sim Na_p^+ - Cl_p^- \qquad (1)$$

On the left hand, then, we have the metabolic component of the acid-base balance and on the right hand the two most important electrolytes. The interaction between the acid-base and the electrolyte balance cannot be more clearly demonstrated.

If you wish to use Base excess instead of Buffer base in (1) it is known from page 8 that $BE_p + 42 = BB_p$ and consequently:

$$\begin{aligned} BE_p + 42 &\sim Na_p^+ - Cl_p^- \text{ or} \\ BE_p &\sim Na_p^+ - Cl_p^- - 42 \end{aligned} \qquad (2)$$

These equations, which correspond to the simplified Gamblegram on page 80, are valid only if the concentration of the residual anions is not increased as it in the following clinical conditions:

1. Ketosis (caused by starvation or diabetes)
2. Lactate increase (usually caused by insufficient oxygen supply to the tissues)
3. Salicylate or methyl alcohol intoxication
4. Kidney insufficiency

One further reservation is needed before the equations (1) or (2) or the simplified Gamblegram is employed. It is assumed that Buffer base was normal, i.e. 42 mmol/litre, when the patient had no acid-base disturbance. Now Buffer base, as stated, is $HCO_3^- + Prot^-$. Sometimes the protein concentration is reduced as in nephrosis. Then Buffer base is correspondingly diminished. A BB value of 33 mmol/litre apparently represents a metabolic acidosis, but may be entirely normal if the protein concentration is reduced from 72 to 36 g/litre. For details see Rooth & Thalme (59).

Although Buffer base varies with the protein concentration, this is not the case with Base excess, which remains normal when there is no metabolic disturbance, as is seen from: Base excess = Buffer base—normal Buffer base.

In the treatment of patients with disturbances showing increased residual anions, it may sometimes be useful to calculate the residual anion concentration and to establish to what extent this is reduced as the result of therapy. Some authors use the expression "anion gap" instead of "residual anions". Anion gap is $Na_p^+ + K_p^+ - (Cl_p^- + HCO_{3\,p}^-) = Prot_p^- + \text{residual anions} - (Mg^2{}_p^+ + Ca^2{}_p^+)$.

The residual anions are obtained, as may be seen from Fig. 11, by substracting $(Cl_p^- + BB_p^-)$ from $(Na_p^+ + K_p^+ + Ca^2{}_p^+ + Mg^2{}_p^+)$. Unfortunately the result may be unreliable even if the original measurements are good, as the error may be large when two more or less equal sums are substracted. This is illustrated in a few examples:

Suppose that the error in the measurement of the cations is 4 percent and that the errors in the Cl_p^- are 3 percent and BB 5 percent. If the errors in the cations and in the anions happen to go in different directions, the following two results may be obtained:

Residual anions $= 153 + 4\% - [(101-3\%) + (42-5\%)] = 159 - 137 = 22$ mmol/l or

$$153 - 4\% - [(101+3\%) + (42+5\%)] = 147 - 148 = -1$$
mmol/l

The examples show that some care should be used in interpreting residual anion changes. If important for the understanding of the treatment, it may be better to measure the individual anions directly if given the technical resources. For practical purposes it suffices to analyze 3-hydroxybutyrate in diabetic ketosis, lactate in tissue hypoxia, and phosphate or sulphate in kidney insufficiency. Sometimes one is faced with an apparently unexplained metabolic acidosis. If the above mentioned anions are analyzed, one will virtually always discover the genesis of the acidosis.

If a reliable figure is needed for the residual anions, it it best to determine Buffer base directly from bicarbonate and protein analyses. $HCO_3^-{}_p$ is read from the Siggaard-Andersen nomogram in Fig. 13 after measurements of P_{CO_2} and pH. $Prot_p^-$ is calculated from the chemical determination of protein using the formula of van Slyke, Hastings, Hiller & Sendroy (40):

$$Prot_p^- = 0.104 (Prot\ g/l)\ (pH - 5.08) \qquad (1)$$

Intracellular electrolytes

The intracellular electrolyte concentration is shown in Table 1. Gamble (6) gave Na_c^+ as 10, K_c^+ as 150, and Mg^{2+}_c as 8—20 meq/litre. Newer measurements, again on muscles, give somewhat higher values for Na_c^+ and K_c^+ (103). Table 1 lists values between those of the above authors.

Table 1
Intracellular and plasma electrolyte concentration (meq/litre)

	Cations			Anions		
	Na^+	K^+	Mg^{2+}	HCO_3^-	$Prot^-$	$Phosphate^-$
Intracellular fluid	15	160	30	10	40	150
Plasma	143	5	2	25	17	1–2

It is not difficult to analyze the cations in the muscles, but measurements of the intracellular anions can be uncertain. In practice the cations are determined and also HCO_3^-. Using the law of electro-neutrality, the sum total of anions is known, of course, but how much is intracellular protein and phosphate, respectively, remains uncertain. It follows that the Gamblegrams drawn from intercellular ion concentrations look better than they are and I have therefore made only Table 1, which should also be taken with reserve.

There is no doubt that intracellularly there are more meq/litre than extracellularly. However, this gives no cause to believe that the osmolarity is different. The osmolarity depends upon the molar concentrations, not upon meq/litre. For instance, $Mg^2_c{}^+$ is about 30 meq/litre, but 15 mmol/litre as Mg^{2+} is bivalent. It follows that the intracellular osmolarity is at least 15 mOsm/litre less than the total cation concentration in meq/litre indicates.

For the measurement of the intracellular cations, concentrated HNO_3 is usually added in order to break down the protein. It is then not possible to distinguish between those cations which were free and osmotic-active and those which were possibly bound to other structures. However, the work of Edelman (27) and others has convincingly shown that the osmolarity is the same in all the body compartments. It therefore seems that, by conventional methods, we overestimate the free cations, and as the sum of anions is calculated as equal, these also reach too high a figure.

It should also be remembered that the intracellular electrolyte concentration is always given per litre of water, whereas the plasma concentration refers to litre of plasma, instead of the corresponding frame of reference, i.e. plasma water. Water is about 93 % of the content of plasma. The corresponding plasma water cation concentration is consequently $153/0.93 = 164$ meq/litre plasma water.

CHAPTER 8

Buffering and Base excess

As already mentioned, the optimal function of the cells depends upon a normal hydrogen ion concentration. The body has two mechanisms to defend itself against too large pH changes:

1 Buffering
2 Compensation usually via the other component, i.e. respiratory compensation in metabolic disturbance, and vice versa. Sometimes, as in diabetic ketosis, there is both a respiratory compensation and increased urinary NH_4^+ elimination.

The buffering is more or less instantaneous. If, in spite of the buffering, pH is sufficiently deranged, the metabolic or respiratory compensation begins, but this is always a relatively slow process. The respiratory compensation may develope over a period of several hours and the metabolic compensation over a period of several days. This is discussed in Chapter 10.

Even if there are several buffer systems, the bicarbonate and protein buffers are the most important. When H^+ is added, the following reactions occur:

$$H^+ + HCO_3^- \rightleftharpoons H_2CO_3 \rightleftharpoons H_2O + CO_2 \qquad (1)$$
$$H^+ + Prot^- \rightleftharpoons HProt \qquad (2)$$

As already mentioned under Terminology and on page 31, H^+ may be added as CO_2. It is then called respiratory and will represent a reaction in (1) going from right to left. Any other addition of H^+ is called metabolic and will drive both equations (1) and (2) to the right.

In respiratory insufficiency, CO_2 is retained and equation (1), as just stated, moves from right to left. The increase in H^+ which then occurs will drive equation (2) as much to the right as equation (1) is driven to the left. It follows that during respiratory acidosis the increase in HCO_3^- in mmol/litre is exactly as great as the decrease in $Prot^-$. Consequently the sum HCO_3^- + $+ Prot^-$ remains unchanged during respiratory changes. This sum = Buffer base. HCO_3^- changes during respiratory acid-base variations, but Buffer base is unaffected due to the combined effect of the two buffer systems. Base excess then also remains unaffected by respiratory changes as:

BE = Actual BB — Normal BB

Normal BB stands for the BB the patient would have if he had no metabolic acid-base disturbance.

The fact that the bicarbonate concentration increases in respiratory acidosis and decreases in respiratory alkalosis, as seen from (1), explains why bicarbonate is an unsuitable parameter for the metabolic component of the acid-base status. Today, when we have Base excess which is unaffected by respiratory changes, and which is obtained without extra analyses, HCO_3^- has lost its historical position as an optimal parameter of the metabolic component.

Singer and Hastings (67) were the first to introduce the concept of Base excess, but they called it ΔBB, i.e. as above BB—NBB. The Astrup group gave ΔBB the name "Base excess" and have later defined it as:
the amount of base or acid needed to bring the pH of the blood back to normal at Pco_2 40 mm Hg and 37°C.

In metabolic acidosis it is an advantage not to work with negative signs; instead of "Base excess" the term "Base deficit" it used. The relation between Base deficit and Buffer base is:

Base deficit = NBB — Actual BB
If there is no metabolic acid-base disturbance:
NBB = BB and both BD and BE = 0

As Buffer base consists of $HCO_3^- + Prot^-$, BB, as already mentioned, varies with the protein concentration. Normally the plasma proteins are 72 g/litre which gives a BB_p of 42 mmol/litre. In whole blood the situation is somewhat different: the red cells contain more protein than plasma and whole blood therefore has a greater buffer capacity. The latter will of course depend upon the haemoglobin concentration (see also Fig. 9 page 26). At a haemoglobin of 150 g/litre (in future, both Hb and glucose will be given as g/litre and no longer as g/100 ml, although the latter is usually retained here) NBB_b is 46 mmol/litre and at Hb 70 g/litre NBB_b is 44 mmol/litre. Base excess, be it BE_p or BE_b, is always 0 when there is no metabolic acid-base disturbance.

The buffer capacity for hydrogen ions added as CO_2 is illustrated in the slope of the lines in Fig. 9. The steeper the slope, the less pH changes for any given change in Pco_2.

If a blood sample with Hb 70 g/litre is equilibrated with gases of different

Pco_2, the slope of the buffer line goes from $BE=0$ to $BB=46$ mmol/litre. If plasma, which of course has $Hb=0$, is equilibrated with gases having varying Pco_2 values, the buffer line goes from $BE=0$ to $BB=42$ mmol/litre. This was also indicated in Fig. 9.

If, instead, a patient now inhales an air/CO_2 mixture so that his Pco_2 increases, or if a patient has a respiratory insufficiency causing his Pco_2 level to rise, it will be observed that his buffer line, established by two or more sets of pH and Pco_2 measurements, follows the slope of neither whole blood nor plasma. Instead, the slope of this "in vivo" buffer line goes from $BE=0$ to $BB=44$ mol/litre or, expressed differently, from $BE=0$ to Hb 50 g/litre in the Hb scale placed under the BB line in fig. 12, page 37.

This buffer line then represents the true "in vivo" buffering. The steeper slope in whole blood is the "in vitro" buffer capacity of the whole blood as observed with the Astrup equilibration technique. The "in vivo" curve may also be called a whole body curve as opposed to the whole blood curve.

The reason for obtaining a slope of the "in vivo" buffer line corresponding to Hb 50 g/litre is the following: The CO_2 increase is distributed over the whole of the extracellular fluid, i.e. partially in the blood, with the buffer capacity corresponding to BB 47 mmol/l, and partially in the interstitial fluid, which is similar to plasma but lacks its proteins. Instead it has a HCO_3^- concentration 1.05 times higher than that of plasma, 25 instead of 24 mmol/litre. This then is BB of extracellular fluid. The net result of the buffering of CO_2 in the total extracellular fluid depends upon the ratio blood/interstitial fluid. It goes without saying that the buffer capacity of the whole extracellular fluid must be less than that of whole blood. As whole blood is now about 1/3 of the total extracellular fluid, it can be calculated that the buffer capacity of the latter will correspond to about 1/3 of the Hb value. This is, in fact, what both Siggaard-Andersen (18) and Schwartz and Relman (62) observed.

One of the advantages of Base excess is that it gives directly, in mmol/litre, the surplus of base or acid present in the blood, or, expressed in another way: Base excess or Base deficit shows directly how many mmol/litre of acid or base must be added in order to normalize the pH of the blood. This information is of great clinical importance. But as the electrolyte infusions given are distributed in the whole extracellular fluid compartment, it is more important to calculate the BE_{ECF} (extracellular fluid) than the BE_b. BE_{ECF} is obtained from the "in vivo" buffer line, either by repeated measurements in one and the same patient with varying Pco_2 or from the BE scale at Hb 5 g/100 ml of blood in the Siggaard-Andersen linear nomogram (page 49 and 50).

There are some further technical advantages in using BE_{ECF}. The BE scale is made after addition of acid and base to blood which is later equilibrated according to the Astrup technique. "In vivo", after an increase in Pco_2, a movement occurs of HCO_3^- out from the red cells into plasma and further to the intersitital fluid. "In vitro", during the Astrup measurements this exchange does not take place because the HCO_3^- generated in the red cells can migrate only into the plasma. Loss of HCO_3^- from the whole blood corresponds to a metabolic alkalosis. In comparison to the "in vitro" analyses, there is a decrease in Base excess "in vivo". The BE_b scale therefore gives too high values at increased Pco_2 levels. (See also Winter et al 21).

If there is a pure respiratory acidosis, say that Pco_2 increases from 40 to 60 mm Hg, Base excess should not change at all. However, because of the way the BE scale was made, there is a fall in BE of about 3 mmol/litre. If the blood volume, particularly the red cell volume, is large in relation to the interstitial fluid volume, this error becomes large. This is the situation in newborn infants. By consistently using BE_{ECF} instead of BE_b, this artefact is avoided, even in dealing with newborn infants (58). There was much confusion for a time over this "in vivo" "in vitro" problem. However, BE_{ECF} has turned out to be a simple and reliable solution.

Another advantage with BE_{ECF} is that it suffices with two measurements, for instance pH and Pco_2, in order to obtain BE_{ECF}, as it is always read on the BE Hb 5 g/100 ml of blood scale. To obtain BE_b the Hb concentration must be known.

Finally the nomogram becomes much more convenient if BE_{ECF} is used, and there is less risk of faulty readings. Fig. 14 shows the simplified Siggaard-Andersen nomogram. Besides the BE_{ECF} scale, only BE_p is retained. BE_p is of course needed in comparing the metabolic component of the acid-base balance with the electrolyte balance, as when using the equation:

$$BE_p + 42 \sim Na_p^+ - Cl_p^-.$$

CHAPTER 9

Calculation of acid-base parameters

Definitions

We repeat from the chapter on terminology:

Respiratory acidosis = P_{CO_2} higher than 40 mm Hg (5.3 kPa)

Respiratory alkalosis = P_{CO_2} lower than 40 mm Hg (5.3 kPa)

Metabolic acidosis = BE values lower than 0 mmol/litre = BD

Metabolic alkalosis = BE values higher than 0 mmol/litre

The pH value which results from the interaction of the metabolic and respiratory components is high if it is over 7.40 ($H^+ < 40$ nanomol/litre) and low if it is less than 7.40 ($H^+ > 40$ nanomol/litre).

According to the recommendation of an ad hoc committe on acid-base terminology, the terms "respiratory acidosis" etc should be used only when these conditions are the primary causes of disease. Thus we have, for instance, "respiratory acidosis in pulmonary insufficiency" and "metabolic acidosis in diabetic ketosis".

The compensations should, for instance *not* be called "respiratory alkalosis in diabetic ketosis". The correct expression should be: "The patient had a metabolic acidosis and, as a compensation, hyperventilation", or, alternatively: "as a compensation, hypocapnia", or, best: "as a compensation, P_{CO_2} was decreased". Correspondingly, in dealing with a case of respiratory acidosis the metabolic compensation is expressed: "BE was increased". Sometimes the word "hyperbaseosis" is also used.

The Siggaard-Andersen curve nomogram

In measuring with the Astrup equilibration technique, whereby the buffer line is defined from two pH readings at different but known P_{CO_2} values and from one anaerobic pH reading, the results are best plotted in a Siggaard-Andersen curve nomogram as in Fig. 12 page 37.

Example: (See again fig. 12). A patient with poliomyelitis and respiratory insufficiency had the following values:

pH = 7.06 anaerobically (H^+ = 87 nanomol/litre)
pH = 7.19 (H^+ = 65 nanomol/litre) after equilibration with Pco_2 30 mm Hg (4.0 kPa).
pH = 7.07 (H^+ = 85 nanomol/litre) after equilibration with Pco_2 55 mm Hg (7.3 kPa).

Why is pH so low? It will be seen at once from the cross, which represents the actual Pco_2 values, that it is above the Pco_2 40 line and we know clinically that the patient has a respiratory insufficiency. She has then a respiratory acidosis. It remains to determine quantitatively what metabolic disturbance there may be. BE_b is read where the buffer line crosses the BE line and is found to be —16 mmol/litre. Thus the patient has a respiratory acidosis and a Base deficit$_b$ of 16 mmol/litre. Both components have moved in the same, acid, direction and each has therefore contributed to the decrease of pH.

In order to obtain BE_{ECF}, or rather in this case BD_{ECF}, another step is needed. If the pH value 7.06 and the Pco_2 value which we have just obtained, i.e. 58 mm Hg, are entered into Fig. 14 BD_{ECF} is read as 13.5 mmol/litre.

In dealing with a case of respiratory insufficiency and acidosis some compensation might be effected, which of course would be concomitant with an elevation in Base excess. However, in this case the patient had a metabolic acidosis over and above her respiratory acidosis. The reason is that the patient, who had difficulty with the expiration of adequate amounts of CO_2, also had difficulty with the inhalation of adequate amounts of oxygen, and had therefore contracted a tissue hypoxia.

We call this type of acid-base balance mixed, i.e. there are two different primary acid-base disturbances. In clinical practice one seldom finds pure respiratory or pure metabolic acid-base abnormalities. Usually there is a metabolic disturbance with respiratory compensation, or vice versa. Sometimes there are, as in the present case, two primary disturbances which affect pH in the same direction; sometimes the situation is even more complicated. (See the results of primary respiratory insufficiency + metabolic compensation + metabolic alkalosis due to diuretics page 71).

Remember that a laboratory should not be called upon to determine what the primary disturbance may be. That answer is obtainable from the clinical picture.

To summarize the curve nomogram: Each pH value which is in the upper half of the diagram represents a respiratory acidosis if the patient has a disease associated with pulmonary insufficiency; otherwise the patient has a respiratory compensation for a metabolic alkalosis.

With the same reservations, an actual pH value below the Pco_2 40 line represents a respiratory alkalosis, alternatively a respiratory compensation for a metabolic acidosis.

Every buffer line which crosses the Base excess curve on the positive side represents a metabolic alkalosis, if the patient has a primary disease which leads to this; otherwise it is a metabolic compensation for a respiratory acidosis. Finally, if the buffer line crosses the Base excess curve on the negative side, the patient has a metabolic acidosis or a metabolic compensation for a respiratory alkalosis.

Respiratory changes are represented by points moving along the buffer line, metabolic changes, by the displacement of the buffer line across the Base excess scale. The influence which the individual components of the acid-base balance have on pH may be read on the pH scale.

The Siggaard-Andersen alignment nomogram

(See Fig. 13). Such a nomogram is sometimes called d'Octagne's nomogram after the French mathematician who first constructed it. It is basically the same nomogram as that introduced by Peters and van Slyke (11) in medicine for the calculation of pH and Pco_2, and later used by Singer and Hastings (64) for the measurement of Buffer base. With regard to the pH, Pco_2 and HCO_3^- scales, these are nothing but a graphic representation of the Henderson-Hasselbalch equation:

$$pH = pK' + \log \frac{HCO_3^-}{S \times Pco_2}$$

where pK' is, for most purposes, a constant = 6.1 and S = 0.03, the solubility coefficient for CO_2.

As we have log $HCO_3^-/S \times Pco_2$ the Pco_2 and bicarbonate scales are logarithmic. The Total CO_2 scale to the left of the nomogram is needed if the CO_2 content of plasma is measured gasometrically. The Base excess scale is new and added by Siggaard-Andersen. As already described on page 44 this scale suffices in clinical practice for measuring BE_{ECF} This makes it possible to simplify the nomogram, as shown in Fig. 14.

In many laboratories today Pco_2 is measured with the electrode. Using this and a pH meter, the following results were noted in a patient with respiratory insufficiency (Fig. 14):

pH = 7.33 (H^+ 47 namomol/litre)
Pco_2 = 71 mm Hg (9.5 kPa)

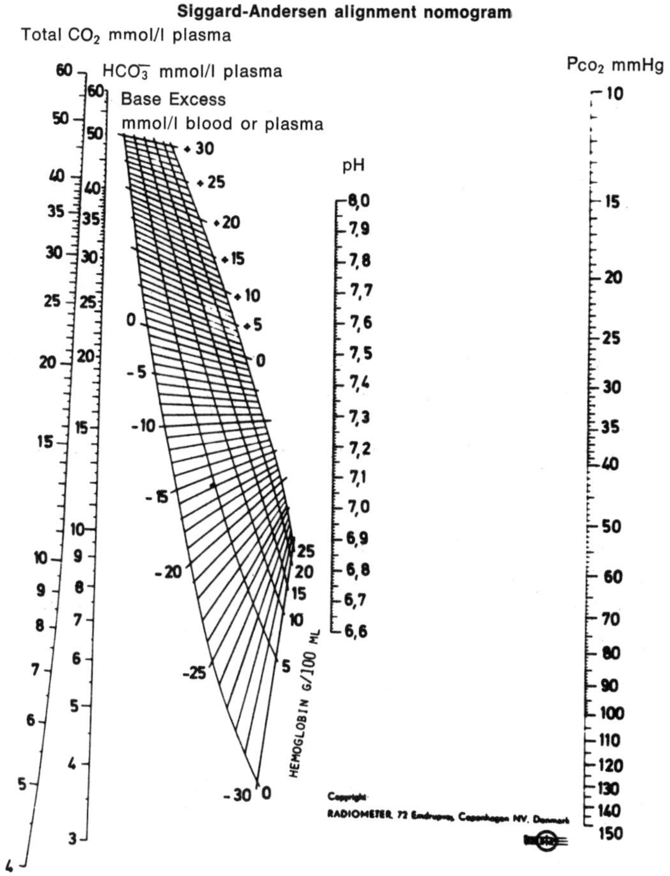

Fig. 13. The Siggaard-Andersen alignment nomogram.
According to our experience, the use of this nomogram is the quickest and simplest way to calculate the different acid-base parameters. It may be used in various ways, depending upon what primary measurements you have and what you want to know.
A special advantage of the nomogram is that it makes it possible to "translate" from one acid-base terminology to another, as will be shown in the subsequent discussion.

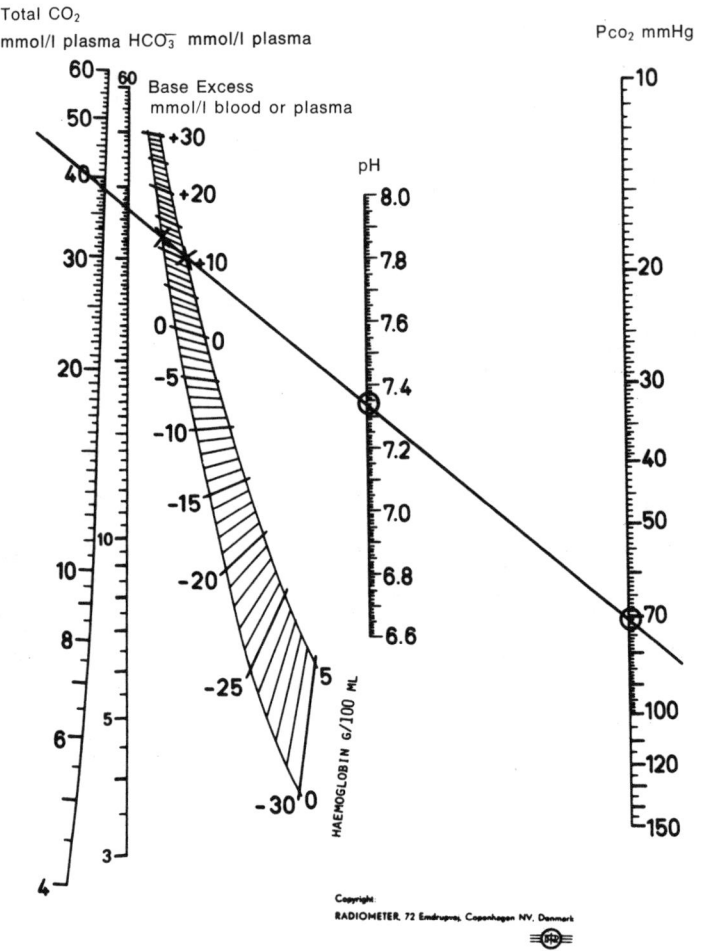

Fig. 14. Acid-base data from a patient with respiratory insufficiency.
By uniting two known points on different scales with a straight line, all the other acid-base parameters are obtained. In the present case, pH and Pco_2 were analyzed and the results marked by circles. BE_{ECF} is read where the line crosses the BE scale at Hb 5, and BE_p at Hb 0. HCO_3^- and total CO_2 in plasma is obtained from the two scales to the left. (See page 48.)

The clinical picture and the direct measurement show that pH is low and that this is caused by a respiratory acidosis. It remains to establish the metabolic component. After having united P_{CO_2} 71 mm Hg with pH 7.33, a straight line is drawn which crosses the BE Hb 5 scale at BE = 10.5 mmol/litre. This patient thus has a respiratory acidosis with a metabolic compensation. The degree of metabolic compensation should always be compared to the expected compensation according to Fig. 21, page 65. It will then be seen that the BE value of 10.5 mmol/litre in our patient falls well within the shaded area in Fig. 21 and that the patient is consequently fully compensated.

If you wish to compare the electrolyte and the acid-base values, BE_p must be used; in Fig. 14 this is read as 12.0 mmol/litre. As normal Buffer base is 42 mmol/litre, BB in this case is $12 + 42 = 54$ mmol/litre. This, then, is the BB value to enter among the anions in a Gamblegram.

The following example is from a patient with diabetic ketosis. In this case total CO_2 in plasma was analyzed gasometrically and pH was measured. pH = 6.96 (H^+ 110 nanomol/litre)
Total CO_2 in plasma = 5.0 mmol/litre

We want to know the degree of metabolic acidosis of this patient expressed in our usual units, i.e. Base excess and P_{CO_2}, in order to obtain the degree of her respiratory compensation. Fig. 15 shows how total CO_2 in plasma 5.0 and pH 6.96 are connected. BD_{ECF} turns out to be 25.5 mmol/litre and P_{CO_2} 21 mm Hg (2.8 kPa).

If the patient had not had hyperventilation, her pH would of course have been lower. What the value would have been we can ascertain by uniting BD_{ECF} 25.5 with P_{CO_2} 40 mm Hg. The value is found to be 6.83. A patient can hardly be expected to survive such a low pH. Note also that this pH value is Standard pH, i.e. the pH value caused by the metabolic component itself when the respiratory component is normal.

The third example is from a man doing intensive exercise at high altitude. Measured values:

pH = 7.14 (H^+ = 72 nanomol/litre)
$HCO_3^-{}_p$ = 13.5 mmol/litre

We want to know why pH is low. Is it because of a respiratory insufficiency with high P_{CO_2} or because of metabolic acidosis? Fig. 16 shows that after uniting pH 7.14 and HCO_3^- 13.5 with a straight line, this crosses P_{CO_2} at 42 mm Hg (5.5 kPa) and BD_{ECF} at 13.5 mmol/litre. As P_{CO_2} was so close to 40 mm Hg, the patient could have no respiratory acid-base disturbance; the pH decrease is therefore due to his metabolic acidosis. For clinical reasons we could suspect that the patient had tissue hypoxia. An expression more in

Siggard-Andersen alignment nomogram

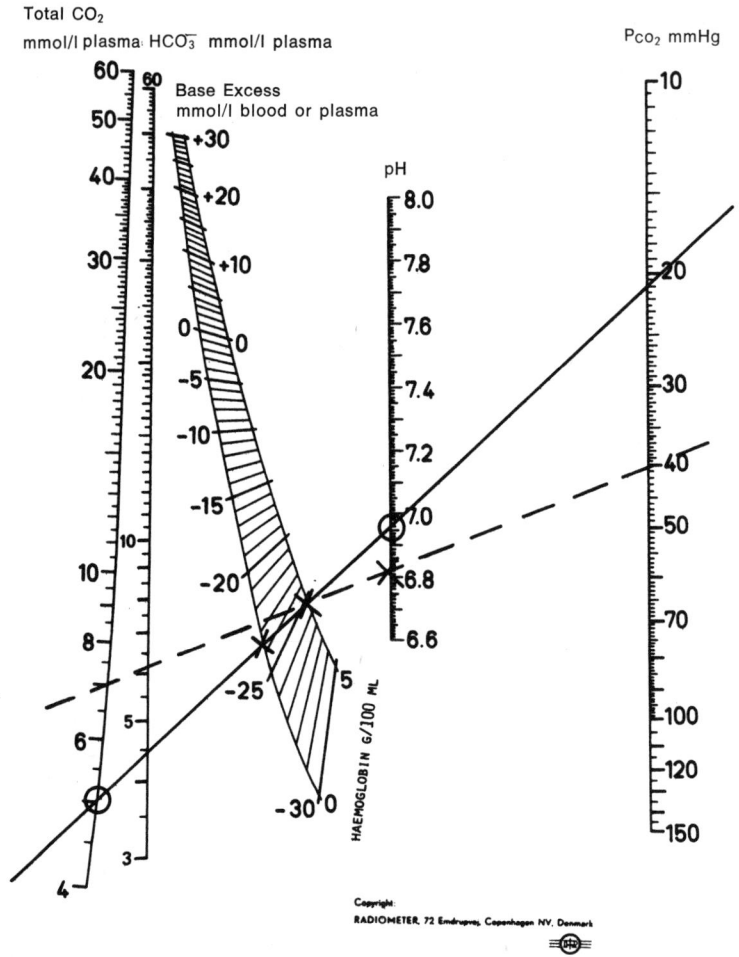

Fig. 15. Acid-base values from a case with diabetic ketosis:
pH and total CO_2 were measured. BE_{ECF} is read on the right hand side of the BE scale and BE_p on the left.
The dashed line illustrates the effect on pH of the hyperventilation. With normal ventilation, i.e. $Pco_2 = 40$ mm Hp, the patient would have had a pH of 6.83. The latter pH is also called Standard pH, or pH_{NR} (non-respiratory).

Siggard-Andersen alignment nomogram

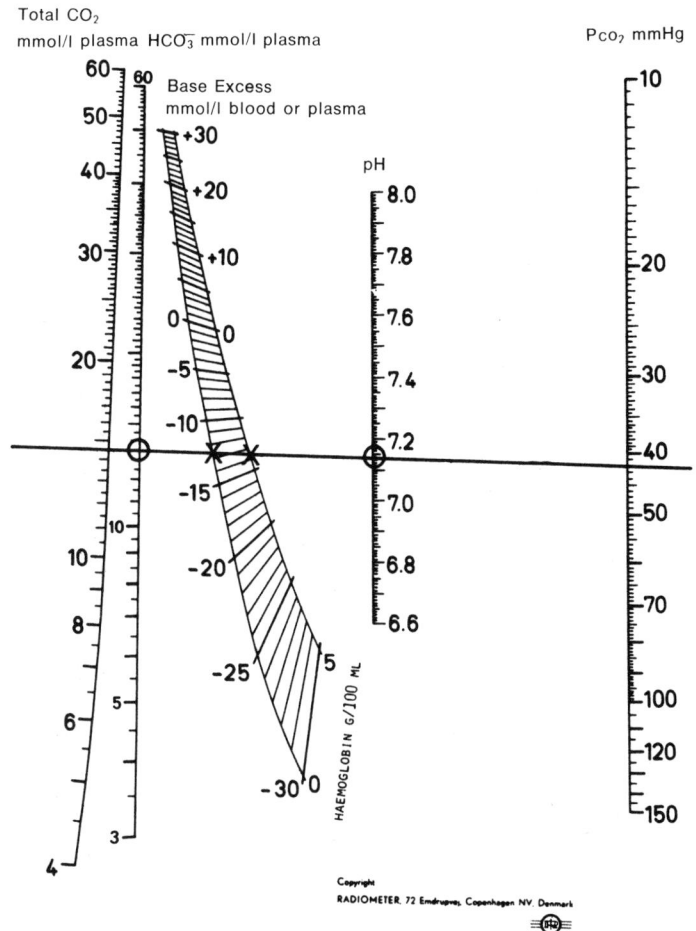

Fig. 16. Acid-base values from a person exercising at high altitude: pH and plasma bicarbonate were determined. BE_{ECF} is read on the right side of the BE scale och Pco_2 on the extreme right. (See page 51.)

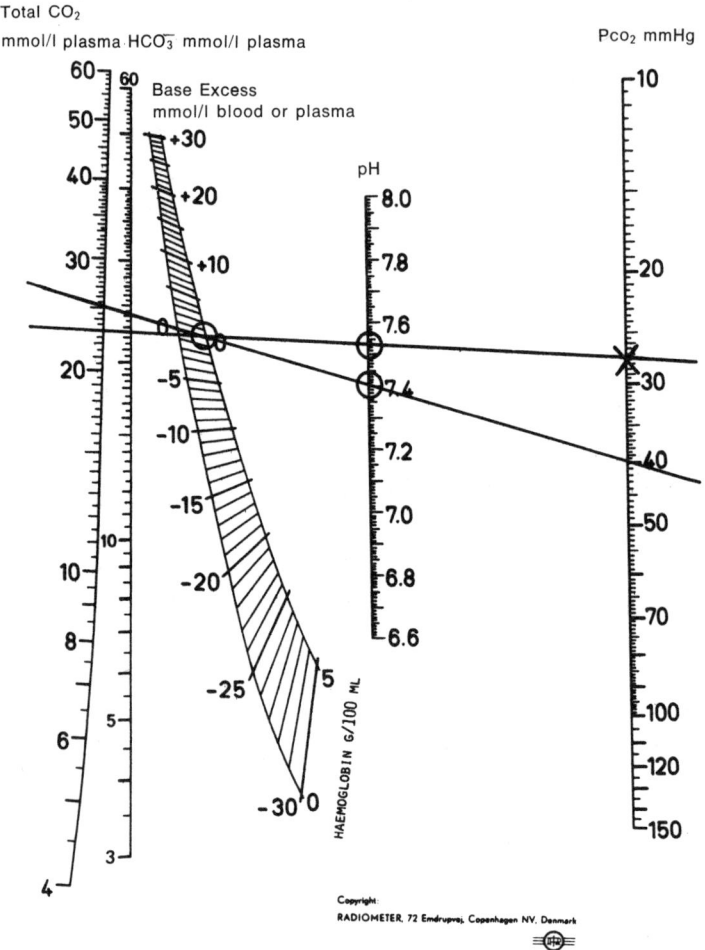

Fig. 17. Acid-base data from a patient with salicylate poisoning:
pH was first measured anaerobically, subsequently after tonometry with a gas mixture with $Pco_2 = 40$ mm Hg. Line 1 connects Pco_2 40 with the pH value obtained after equilibration. From this line BE_{ECF} is read. Finally, line 2 is drawn from BE_{ECF} to the actual pH and over to the Pco_2 scale.

use some years ago was "oxygen' debt". Chemically, an increase in lactate would undoubtedly have been shown if that part of the residual anions had been measured.

The fourth example is from a patient with acute salicylate poisoning. (See Fig. 17). Measured values:

pH = 7.53 (H^+ = 28 nanomol/litre)
pH = 7.40 after equilibration of whole blood with a gas mixture having a Pco_2 of 40 mm Hg (5.3 kPa).

This type of measurement may be made with comparatively simple equipment. In a similar manner Saling (14) makes his measurements according to the technique of Astrup (20). It seems possible the in the future smaller hospitals will perform acid-base analysis in this way, whereas Po_2 and Pco_2 determinations with electrodes will be reserved for the larger laboratories.

Whether the curve nomogram or the alignment nomogram is used, the calculation is somewhat laborious. At first a line no 1 is drawn as in Fig. 17, uniting Pco_2 40 mm Hg and pH 7.40, i.e. the result of the pH measurement after tonometry. This line gives BE_{ECF} = 0 mmol/litre. Finally the BE_{ECF} value is connected to the actual pH value and Pco_2 is read as 28 mm Hg.

Estimation of acid-base status from inadequate measurements

This nomogram is useful even when we have less measurements than are needed for a complete analysis of the acid-base balance. Take again a case of diabetic ketosis with the following values:
pH = 7.20 (H^+ = 63 nanomol/litre)
It was noted that the patient hyperventilated i.e. had Kussmaul's deep breathing. How severe is the metabolic acidosis of the patient?

Of course we would have needed a Pco_2 value but fortunately we have semiquantitative information from the clinical observation that the patient hyperventilated. In such cases Pco_2 can hardly be higher than 30 mm Hg (4.0 kPa) and is probably not much below 20 mm Hg (2.7 kPa). If, then, as in Fig. 18, these two Pco_2 values are connected with lines going through the actual pH value of 7.20, BD_{ECF} is found to be 19 mmol/litre (at Pco_2 20 mm Hg) and 15 mmol/litre (at Pco_2 30 mm Hg). The therapy will be the same whether the Base deficit is 15 or 19 mmol/litre, and in spite of the insufficient data, we have obtained enough information for our clinical needs. It would also have been possible, of course, merely to discuss the pH value as such and to state that it was not very low and therefore the metabolic acidosis was not too pronounced. It is probable however, that most doctors would have underestimated the degree of metabolic acidosis if they looked only at the pH value. The combination of the pH value and estimated Pco_2 levels gives a safer basis than pH alone.

The slide rule of Severinghaus

Severinghaus constructed a slide rule for the calculation of the acid-base parameters. It is based on the Siggaard-Andersen alignment nomogram. Some clinician prefer the nomogram, others the slide rule, according to whichever method is habitually used.

The slide rule has in addition scales for the temperature correction of the acid-base parameters and for the alveolar equation and is therefore most useful for the physiologist studying respiration.

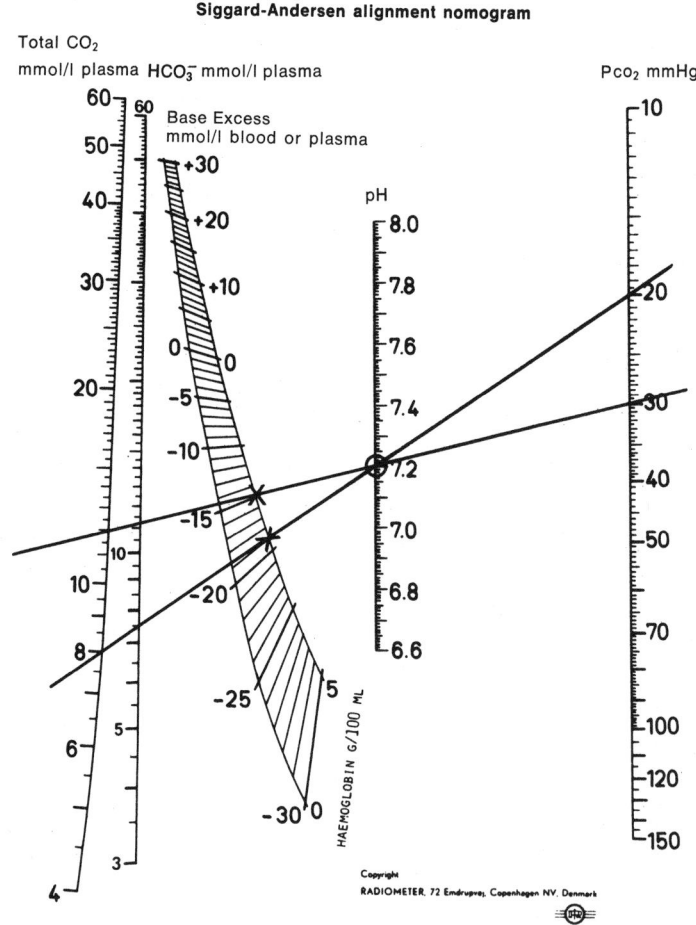

Fig. 18. Acid-base values from a patient with diabetic ketosis:
Only pH was determined. Pco_2 was estimated as not higher than 30 mm Hg and probably not lower than 20 mm Hg. A line uniting the Pco_2 with the actual pH gives BD_{ECF} 19 mmol/litre. If Pco_2 were 30 mm Hg, BD_{ECF} would have been 16 mmol/litre.

BE/Pco₂ diagram

If for different pH values, 7.10, 7.20 etc., Base excess and Pco_2 are read on a Siggaard-Andersen nomogram and the values plotted in a BE/Pco_2 diagram, it will be found that the pH lines are almost straight, as first demonstrated by Winters, Engel, and Dell (18). These plots vary with the haemoglobin concentration and for BE_b a whole set of diagrams would be needed, but, as has already been discussed (page 44), it suffices to use BE_{ECF} and to make one BE/Pco_2 plot at Hb 50 g/litre.

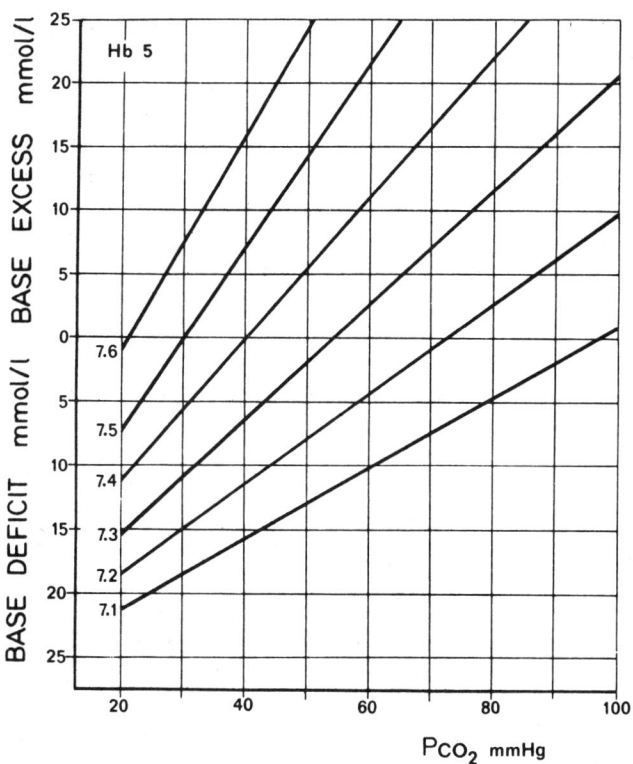

Fig. 19. Base excess/Pco₂ diagram.

Pco₂/BE diagram

According to mathematical tradition, the independent variable should be plotted along the absciss and the dependent along the ordinate. Then if we study how Pco_2 is affected by metabolic changes, a Pco_2/BE diagram should be used, as in Fig. 23 (page 74).

There is no competition between the BE/Pco_2 diagram and the Siggaard-Andersen nomogram; they have different uses. The diagram is invaluable in the following two circumstances:

1. In observing the day-to-day or hour-to-hour changes in one and the same patient, one single point in this plot gives all the three acid-base parameters, pH, Pco_2, and BE. In these circumstances the BE/Pco_2 diagram gives a much better over-all picture of the acid-base changes in the patient than do individual plots or tables of pH, Pco_2, and BE. There is always a risk in focusing on only one of the parameters. For instance, as described on page 67, a correction of Pco_2 until a normal value is reached may, in a case of compensated pulmonary insufficiency, lead to a dangerously high pH value.

2. In evaluating the degree of respiratory compensation for a metabolic disturbance or a metabolic compensation for a respiratory acid-base abnormality.

Using BE/Pco_2 diagrams with inlaid compensation lines, mixed acid-base disturbances may be quantitatively analysed.

Formulas for calculations

The above described calculations of the acid-base parameters are all based on the measurements of Siggaard-Andersen and upon the nomogram he devised. However, it as also possible to calculate Base excess from pH, bicarbonate, or Pco_2 and protein determinations using factors determined mainly by van Slyke and co-workers (40) but also by Siggaard-Andersen (15). The latter has also given the formulas (63):

$$BE_b = (1 - 0.0143 \times Hb) \, [\Delta HCO^-_3 + \Delta pH \, (1.14 \times Prot + 1.4 \times Hb)] \quad (1)$$

BE_{ECF} where $Hb = 5$ g/100 ml becomes:

$$BE_{ECF} = 0.93 \, [\Delta HCO^-_3 + \Delta pH \, (1.14 \times Prot + 7)] \quad (2)$$

and

$$BE_p = \Delta HCO^-_3 \times \Delta pH \times 1.14 \times Prot \quad (3)$$

where ΔHCO^-_3 stands for the deviation in the bicarbonate concentration from the normal value of 24.0 mmol/litre and ΔpH for the deviation from 7.40.

Radiometer A/S recently has manufactured a small computer for the calculation of acid-base parameters. This is of course the most convenient method. Several computer programs have been published. Reference to these will be found in Engleson et al (51).

Corrections for unsaturated haemoglobin

According to the definition, Base excess should be given at the actual oxygen saturation of the blood, which is automatically the case when Base excess is calculated from a pH and a Pco_2 determination.

However, using the Astrup equilibration method, where the blood is always mixed with a gas of high oxygen concentration, any initially unsaturated haemoglobin becomes saturated. As oxyhaemoglobin is a stronger acid than haemoglobin, there is a certain pH change due to this factor after tonometry. For instance, in fetal blood with an oxygen saturation of 40—50 % or less, a correction of BE with 2—3 mmol/litre is needed. In arterial blood from adults no correction is needed, and in practice most authors delete the correction for the unsaturated haemoglobin and unfortunately make no reference to it.

Siggaard-Andersen (15) gave the following correction factor:

BE_b (actual oxygen saturation) = BE_b (100 % oxygen saturation) + 0.3 × × Hb(g/100 ml) × (100 — actual oxygen saturation)/100
If the Hb concentration is 15g/100 ml and the oxygen saturation 60 %,
$BE_b = BE_b$ (after equilibration) + 0.3 × 15 × 40/100 and
$BE_b = BE_b$ (after equilibration) + 1.8 mmol/litre

The factor 0.3 is empirically found and according to other studies, at least Pco_2 is better determined without this correction (34, 45). It therefore seems safe to state that for clinical purposes the correction may be deleted, but that if is needed in research the safest method is to determine the factor in each case.

The use of capillary or venous blood

Occasionally considerable errors are obtained in pH and Pco_2 if capillary or venous blood is used. The reason for this is not, as some may believe, the effect of the unsaturated haemoglobin. Rather, it is that sometimes, particularly, of course, in shock, the peripheral blood flow is slow and the capillary blood is no longer representative of the arterial blood. Capillary blood taken from a finger tip usually has a Pco_2 which is only a few mm Hg higher than that of the arterial blood. If a stasis is applied, considerably higher P_{CO_2} levels may also be found, as during peripheral vasoconstriction. As a rule, Base excess values are reliable also from venous or capillary blood samples, but one always should interpret P_{CO_2} and pH values with due caution. For instance, never place a sick patient in a respirator on a high capillary P_{CO_2} value. In such cases arterial confirmation must always be obtained first.

CHAPTER 10

Respiratory insufficiency

The acid-base and electrolyte changes in patients with pulmonary insufficiency are often complex and difficult to interpret correctly. However, if given the opportunity to follow the evolution of the case, it is usually possible to understand what happens.

Let us, at first, divide the condition of respiratory insufficiency into three stages:

1 Acute insufficiency
2 Acute insufficiency + tissue hypoxia
3 Chronic insufficiency with renal compensation (increased Base excess value). (See pages 64 and 65.)

1 In breath-holding or any other circumstance when ventilation is acutely suspended or considerably reduced, Pco_2 increases and pH falls (H^+ increases). These changes are represented by a point going along the buffer line $A_2—B_2$ in Fig. 7 page 23. As soon as adequate ventilation is restored, the CO_2 surplus i blown off and Pco_2 and pH return to normal values.

2 When respiration is insufficient for CO_2, it is usually still more insufficient for O_2. Po_2 will fall, but the oxygen saturation of the haemoglobin will only be slightly affected by the initial fall in Po_2 because of the S-shape of the oxygen dissociation curve. In spite of a decreased Po_2, an adequate oxygen supply to the tissues is possible for a time. However, should the supply per time unit decrease below the normal oxygen consumption, or should consumption be increased because of muscular effort, the tissues will not receive sufficient oxygen. A hypoxic metabolic acidosis ensues and the lactate concentration increases.

When the metabolism is anaerobic, the only energy generated comes from the breakdown of glucose, but only 1/19 of the amount of energy usually produced now becomes available. Some quantitative data about the rate of lactate production in hypoxia is given in ref (97). (See also the chapter on tissue hypoxia in Rooth (13).

Increase in lactate in a patient with pulmonary insufficiency is not usually called lactoacidosis, although of course chemically this is correct. Lactoacidosis is a term usually reserved for severe and, often, fatal cases with pronounced lactate increase and low pH values. (See page 87.)

When a patient is in acute need of artificial respiration it must be assumed that the oxygen supply has been deficient for a time and that consequently the respiratory acidosis is complicated by a metabolic acidosis, i.e. that the patient has a mixed acidosis.

Clinical example: A boy 13 years of age with acute poliomyelitis had difficulty in breathing on the fourth day of his disease and was given artificial respiration because of the following values given in Table 2:

Table 2

	pH	H^+ nanomol/litre	P_{CO_2} mm Hg	kPa	BE_{ECF} mmol/litre
Initially	7.11	78	72	9.6	6.0
After 10 min. in respirator	7.26	55	58	7.8	1.0
After 1 hour in respirator	7.47	34	30	4.0	1.0

P_{CO_2} is rather high, but the initially low pH value is due to the combination of respiratory and metabolic acidosis. Both parameters affect pH in the acid direction. A few minutes after the patient was given artificial respiration, when he also was given supplementary oxygen, the metabolic acidosis was abolished and P_{CO_2} was beginning to normalize. After one hour a respiratory alkalosis was noted and the artificial ventilation was reduced.

It will be seen that the pH changes were considerable. H^+ was reduced by half in one hour. There are cases on record where the pH changes have been even greater going, for instance, from 7.04 to 7.71. K_p^+ decreased in one case from 6.5 to 2.6 mmol/litre. Thus this case stresses the intimate relationship treatment of a patient with respiratory paralysis, the pH value may in a short time vary from the lowest to the highest values compatible with life.

At the same time the patient underwent the pH changes, his K_p^+ decreased from 6.5 to 2.6 mmol/litre. Thus this case stresses the intimate relationship between pH and K^+, which is too little known. Besides the formula on page 81, one can use that described by Siggaard-Andersen: For a decrease in pH of 0.1 units, K_p^+ increases 0.3 mmol/litre.

In acute respiratory insufficiency there are potassium changes, but no other important electrolyte alternations. The K_p^+ drop may be clinically important, resulting in fall in blood pressure and cardiac disturbances.

In a patient with tissue hypoxia the Buffer base concentration is reduced and the lactate level increased. At the same time there is a certain movement of Na^+ and water into the cells in exchange for H^+.

Renal compensation for respiratory acidosis

If the alveolar hypoventilation is sustained, the lowered pH will gradually be corrected by renal compensation.

According to the Henderson-Hasselbalch equation:

$$pH = pK' + \log HCO_3^-/0.03 \times Pco_2$$

As pK' is 6.1 the equation at pH 7.40 may be rewritten:

$$7.40 = 6.1 + \log HCO_3^-/0.03 \times Pco_2 \text{ and}$$
$$1.30 = \log HCO_3^-/0.03 \times Pco_2$$

taking an antilog of 1.30 we get:

$$20 = HCO_3^-/0.03 \times Pco_2 \text{ and } HCO_3^-/Pco_2 = 0.6$$

It follows that normally the relation between the bicarbonate concentration in plasma and Pco_2 is 0.6. As Pco_2 increases in respiratory acidosis, HCO_3^- also must increase if pH is to be normal. The kidneys achieve this by eliminating H^+ and retaining HCO_3^-, which is reabsorbed. Both H^+ and HCO_3^-, as always, come from the CO_2 produced during metabolism according to Fig. 20.

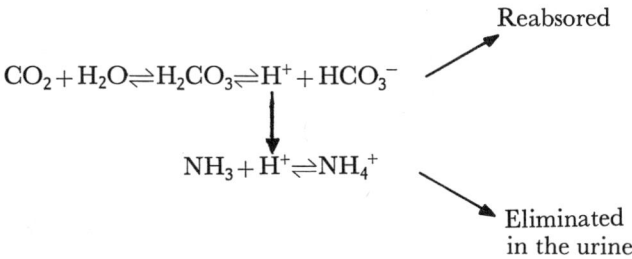

Fig. 20. The H^+ elimination and HCO_3^- conservation of the kidney.
The kidneys cannot eliminate H^+ in higher concentration than about 1 mmol/litre, but about 100 to 500 mmol/litre are needed for adequate compensation. H^+ is therefore bound to NH_3 derived from glutamate and NH_4^+ is eliminated in the necessary quantities.

CHAPTER 11

Metabolic compensation for respiratory acidosis

It was stated above that the acid-base balance was governed by a physiological law according to which the body tries to maintain an extracellular fluid pH of 7.40. This holds true whether the primary disturbance is metabolic or respiratory, and the compensation is brought about by the "other" component of the acid-base balance, i.e. that which is not primarily affected.

The compensation seldom brings pH back to 7.40. The statement, quoted from textbook to textbook, that when fully compensated, pH is brought back to normal, otherwise it is only partially compensated for, is a didactic oversimplification. If compensation always caused pH to return to 7.40, the whole acid-base problem would follow the pH 7.40 line in the BE/$P{co}_2$ diagram in Fig. 21.

Although compensation does not always make pH normal, it can normalize pH in a predictable way, as revealed by several studies during the past decade. This holds true whether the primary disturbance is metabolic or respiratory.

Fig. 21 shows the mean value for the metabolic compensation for respiratory acidosis with its 95 % confidence interval of ± 0.05 pH units. The compensation line appears to be valid in different countries, in young children as well as in adults (8, 50, 67).

Only with the aid of a diagram of this type may the Base excess and pH value be correctly interpreted in a case of respiratory insufficiency.

Some prefer a HCO_3^-/H^+ diagram (50) and Siggaard-Andersen has included both the respiratory and the metabolic compensations in a log $P{co}_2$/pH diagram.

By plotting the actual data of the patient in a diagram as in Fig. 22, the best information about the status of the patient is obtained together with reliable guidance for further treatment.

It should be remembered that it takes about one week for a metabolic compensation to be fully developed. In an acute respiratory insufficiency, say when $P{co}_2$ increases from 40 to 60 mm Hg, Base excess initially remains

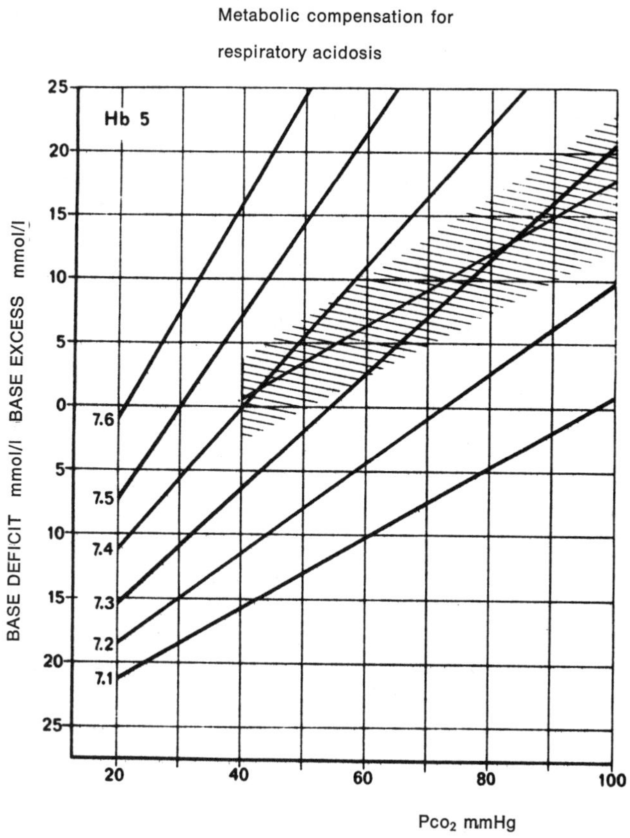

Fig. 21. Base excess/Pco₂ diagram including the metabolic compensation line and its 95 % confidence limits.

unchanged. The initial acid-base status of the patient is represented by point A in Fig. 22 and after his Pco₂ increase, by point B; i.e. there is a horisontal movement along the Pco₂ axis and no change in Base excess; pH has decreased to 7.26. If now the respiratory depression remains at the same level, the Pco₂ remains the same but pH will gradually increase as a result of a rise in Base excess. Within a week a final value corresponding to point C with a pH of 7.34 and a Base excess of 6 mmol/litre is reached.

What pH will a patient with Pco₂ 85 mm Hg have a) initially, b) after full compensation?

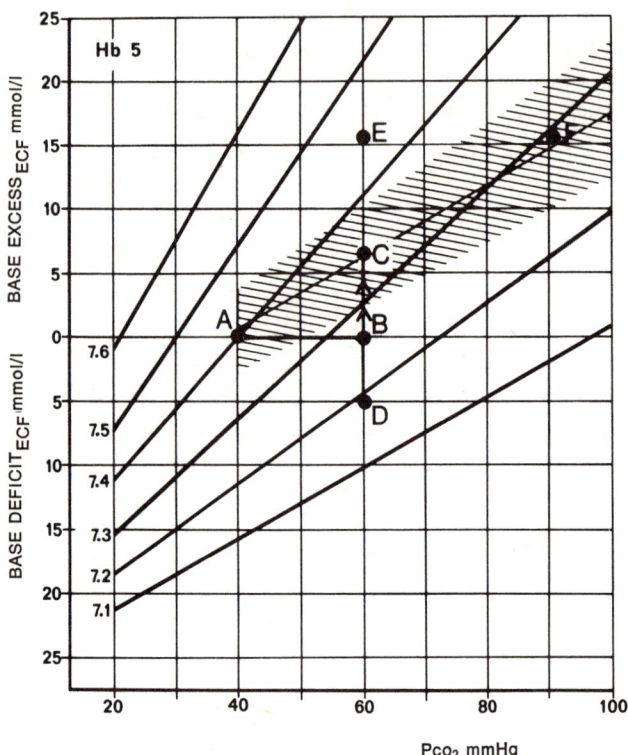

Fig. 22. The metabolic compensation for respiratory acidosis.
The same basic picture as in Fig. 21.
In acute respiratory acidosis the values go from point A to point B, i.e. horisontally. Once the metabolic compensation of the kidneys begins, pH gradually increases toward point C. If the patient suffers acute tissue hypoxia, point D may be reached.

Point E represents the combination of the metabolic compensation and an added iatrogenic metabolic alkalosis caused by diuretics. Point F is the final point which could be reached if there is a respiratory compensation for the iatrogenic alkalosis.

Assuming that there is no initial acid-base disturbance, the answers are: a) 7.15, b) 7.29. a) could have been calculated from the Siggaard-Andersen alignment or curve nomogram, but b) can only be answered by the use of the BE/P_{CO_2} diagram in Fig. 21.

Once we have learned that it is "normal" for a patient with a sustained P_{CO_2} level of 60 mm Hg to have a pH of 7.34 ± 0.05 it follows that if the patient has a significantly higher or lower pH he must have some metabolic disturbance over and above the respiratory acidosis and metabolic compensation. We have here assumed a constant Pco_2 of 60 mm Hg. Any pH variation must be due, then, to a metabolic change.

A higher pH value, say 7.45, indicates that the patient has a respiratory insufficiency + metabolic compensation + metabolic alkalosis.

A lower pH value, say 7.20 (point D in Fig. 22), shows that the patient has a respiratory acidosis + metabolic compensation + metabolic acidosis. The existence of the metabolic acidosis is revealed by the fact that the pH of the patient is lower than it should be from the respiratory acidosis itself. *Quantitatively the metabolic acidosis can only be evaluated if the degree of the previous compensation is known.*

If the compensation developes normally, i.e. if after one week pH drops to 7.20 (point D in Fig. 22), there is a metabolic acidosis of 11 mmol/litre. The metabolic component, Base excess, has changed from +6 to —5 mmol/litre. This, then, is the degree of acidosis and not the 5 mmol/litre Base deficit, as beginners might believe.

As stated above, a pH significantly above the compensation line, say 7.45 at a Pco_2 of 60 mm Hg, signifies:

1. respiratory insufficiency +
2. metabolic compensation +
3. metabolic alkalosis

I can well understand colleagues who think at this stage that the acid-base terminology has become l'art pour l'art. Why bother to differentiate the Base excess into compensation and metabolic alkalosis? 2 plus 3 gives BE 16 mmol/litre as seen from point E, Fig. 22. The explanation is that such mixed acid-base disturbances are common; only by understanding them properly can we treat them correctly.

Patients with respiratory insufficiency often need diuretics. Most of the diuretics induce a metabolic alkalosis (page 107). This is the explanation for 3., which usually is iatrogenic. As this situation is common and may lead to an aggravation of the Pco_2 increase and possibly also to the prolongation of artificial respiration, there is cause to go into the mechanism of the pH changes in some detail.

Sodium retention is often present in chronic pulmonary insufficiency. This does not manifest itself as increased Na_p^+ but the total amount of extracellular sodium and water increases in proportion. (See also page 99.) Initially this may only be revealed as an increase in weight, but manifest oedema will gradually be noted. At the same time the total lung volume is decreased as part of the oedema becomes localized in the lungs where there is little pressure resistance. Lung function will deteriorate further and without the use of diuretics the patient will have entered into a vicious circle.

The sodium retention is probably caused by the fact that the kidneys, as part of the metabolic compensation, eliminate as much as 500 mmol/litre 24 hours of H^+ in the form of NH_4^+. This will bind the available anions of which there are not enough at hand to balance the Na^+ elimination as well.

Whether the sodium-and-water-retaining patient is treated with some of the thiazides or with the fast-acting diuretics Frusemide or Etacrynic acid, a metabolic alkalosis of about 10 mmol/litre develops within a few days.

A patient with Pco_2 60 mm Hg (8.0 pPa) and an optimal compensation has, according to Fig. 22, a Base excess of 6 mmol/litre and a pH of 7.34 (H^+ 46 nanomol/litre). When an iatrogenic metabolic alkalosis induced by a diuretic, is added to this, BE will become $6+10=16$ mmol/litre. This is represented at E in Fig. 22 and pH is 7.45 (H^+ 35 nanomol/litre).

Here we have another example of the value of recognizing the basic disease in evaluating the acid-base data and not simply focusing on the figures given by the laboratory. Our patient had a respiratory acidosis with low pH. However, at this stage it was noted that pH was higher than normal. The patient still had pulmonary insufficiency with high Pco_2 but the fact that the resultant pH was higher than normal can only be explained by a concomitant metabolic alkalosis.

The patient at this stage risks entering a vicious circle and not being correctly treated. As pH is high, the body tends to normalize it. As the pH elevation is due to a metabolic alkalosis, this can be done only by the respiratory component and of course by a further Pco_2 increase. As a result of this, the patient may reach point F in Fig. 22.

If the pH is 7.40, it would be tempting to believe that the patient has a primary respiratory acidosis with a Pco_2 of 67 mm Hg and — according to the old terminology — full compensation. The Pco_2 would then go from 60 to 67 mm Hg during the days of observation, causing a slight deterioration in the respiratory function. Another possibility, but perhaps less likely to occur, is that the pH would reach 7.29 as in point F, Fig. 22. In such a case Pco_2 would have increased from 60 to 91 mm Hg.

Only by knowing the different mechanisms at work can we follow the process and understand how, first, the patient developed metabolic alkalosis as the result of diuretic therapy and, secondly, how his Pco_2 increased in order to normalize the pH. Quantitatively the magnitude of the factors may be sorted out after plotting the data, as in Fig. 22. It goes without saying that the changes are here described schematically, while in clinical practice the changes are not so clearcut and rectilinear as in Fig. 22. Som clinical examples are given in Rooth (13).

To return to Fig. 22, there are two triangles: 1) ABC and 2) CEF.

ABC represents the primary disease and its compensation, while CEF represents the iatrogenic metabolic alkalosis with its compensation.

Before discussing how to deal with the situation, a few words must be said about what not to do.

If the focus is on Pco_2, which is reasonable as the primary disease carries an increased Pco_2, it may be decided that the patient should be placed in a respirator to bring down Pco_2 to 40 mm Hg.

What would then happen? The Base excess value of 16 mmol/litre would of course not be affected by the artifical ventilation and the resulting acid-base data would be

Pco_2 40 mm Hg (5.3 kPa)
BE_{ECF} 16 mmol/litre
pH 7.60 (H^+ 25 nanomol/litre)

This pH value may be obtained from any of the different diagrams since both Pco_2 and BE_{ECF} are known. It follows that such artificial respiration would admittedly normalize Pco_2 but the pH would be dangerously high and K_p^+ would become dangerously low. (See also page 81.) This illustrates the fact that one single acid-base value never should be used individually in order to monitor treatment, but that all three, i.e. pH, Pco_2, and BE should be watched.

Let us return to the treatment of the iatrogenic metabolic alkalosis.

Metabolic alkalosis is always accompanied by low Cl_p^-, and a prerequisite for normalization of the metabolic alkalosis is an increase in the chloride concentration. As chloride is present mainly in the extracellular fluid, Cl^-, in contrast to Na^+ and K^+, may not be mobilized from other body compartments. Remember that chloride in itself does not give the patient acidity as it is no acid (no proton donator); only if Cl^- can be substituted for HCO_3^- will the alkalosis disappear. The necessary concomitant H^+ movements will either take place by H^+ addition, as when NH_4Cl is given, or by HCO_3^- loss by the kidneys.

Theoretically Cl^-_p may be increased in three different ways:

1 by administering NaCl
2 by administering NH_4Cl
3 by administering an acidifying diuretic agent

1 can not be contemplated here as there already was too much Na^+ in the extracellular fluid.
2 In the liver, NH_4Cl is transformed into $NH_3 + H^+$, i.e. the reversed reaction from that which occurs in the kidneys during metabolic compensation (page

63). Ammonium chloride may be given only if the liver is healthy and able to handle the NH_3. NH_4Cl may be given, 1 gram 3 times daily, either in a solution or in coated tablets. When the pH of the patient falls without a concomitant Pco_2 decrease, enough will have been given. This will only happen when pH has fallen below the compensation line in Fig. 21, page 65. 3 Another often more convenient way to restore the acid-base balance is to give Acetazolamide (Diamox®). A similar but less pronounced effect is obtained by administering Triamterene or Spirolactone. The patient will then lose $NaHCO_3$ and water in the urine and Cl^-_p will increase pari passu with the decrease of the extracellular fluid volume.

Clinical example of metabolic compensation for respiratory acidosis

The patient was a woman 56 years of age with emphysema and respiratory insufficiency. Her acid-base and electrolyte status was:

pH 7.37 (H^+ 42 nanomol/litre)
Pco_2 56 mm Hg (7.5. kPa)
BE_{ECF} 6.5 mmol/litre
BE_p 7.5 mmol/litre
Na_p^+ 142 mmol/litre
Cl_p 88 mmol/litre

It is a good exercise to plot these data in a Gamblegram.

Looking first at the pH value of the patient, it will be seen that it is normal. Women have a lower mean Pco_2 than men, 38 versus 40 mm Hg. Judging then from the pH, it is not possible to determine if the patient has a respiratory insufficiency with metabolic compensation or a metabolic alkalosis with respiratory compensation. The rest of the acid-base data offers no further solution to the problem. However, the pH, if anything, is on the low side; it is therefore more likely that we are dealing with a primary acidosis than with an alkalosis. Also, it sometimes happens, for reasons that are not well enough known, that patients with metabolic alkalosis have no, or poorly developed, respiratory compensation.

The deciding factor as always, is the clinical situation. We know that the patient has emphysema and respiratory insufficiency. Consequently the Pco_2 increase is primary and the BE increase compensatory. By plotting Pco_2 56 mm Hg and BE 6.5 mmol/litre in Fig. 21, page 65, we note that the compensation is of the expected order of magnitude.

Clinical example of metabolic compensation for respiratory acidosis + iatrogenic metabolic alkalosis due to diuretics:

This patient was a 53-year old woman with bronchiectasis, pulmonary insufficiency, and secondary polycytaemia of 2—3 years' standing. She was brought to hospital as an emergency case because of haematemesis after having had intermittent epigastric pains for a few days. On admission she had difficulty in breathing, oedema of the legs, and a tender abdomen with the liver three fingers below the costal margin. A tracheotomy was performed and the patient was given intermittent positive pressure, diuretics, and digitalis. The available data are given in Table 3.

By combining total CO_2 45 with pH 7.40 in the Siggaard-Andersen alignment nomogram, Pco_2 71 mm Hg is obtained; at the same time BE_{ECF} may be read as 17.5 mmol/litre and BE_p as 19.0 mmol/litre.

On admission we know only the patient's total CO_2, not her pH, but some estimations are worthwhile. We do know that when she entered hospital her condition was worse than on day 12 when her Pco_2 was 71 mm Hg. Most likely the Pco_2 on day 1 was higher than 71 mm Hg. Let us guess that it was 80 mm Hg, although it could very well have been higher still. Pco_2 80 mm Hg and total CO_2 33.8 mmol/litre give a BE_{ECF} of 4.5 mmol/litre. The high total CO_2, then, was due mainly to the high Pco_2 and not to a pronounced metabolic alkalosis. According to Fig. 21, page 65, we would have expected a BE_{ECF} of about 11 mmol/litre and the BE_{ECF} value of 4.5 which we obtained is below the 95 % confidence limit. This indicates that the patient had a certain amount of tissue hypoxia at that time.

On day 12 the patient had a pH of 7.40 above the compensation line in Fig. 21 and not only a metabolic compensation but in addition a metabolic alkalosis. The latter was caused by diuretic therapy because of oedema and an enlarged liver.

Table 3

	day 1	day 12
pH (H^+ nanomol/litre)		7.40 (40)
Oxygen saturation (So_2) percent		100
Pco_2 mm Hg (kPa)		71 (9.6)
Total CO_2 mmol/litre	33.8	45.0
Na_p^+ mmol/litre	143	136
Cl_p^- mmol/litre	108	71
K_p^+ mmol/litre	5.0	2.8
Haemoglobin g/litre		190

This is again a case in which it is clear that it is not enough to look at only one of the parameters in the acid-base balance. Here we found a pH of 7.40 and could easily have misjudged the situation if we had not had a compensation line as in Fig. 21. At Pco_2 71 mm Hg the expected optimal compensation is about 9 mmol/litre, but the observed Base excess value was 17.5 mmol/litre. The difference of 8.5 mmol/litre may be attributed to the diuretic therapy.

CHAPTER 12

Metabolic compensation for respiratory alkalosis

Sustained hyperventilation is so rare that it is not necessary to discuss, in too much detail, the metabolic compensation for respiratory alkalosis.

Chronic hyperventilation occurs mainly in pregnant women and in people living at high altitudes. As stated above the mean Pco_2 in women is 38 mm Hg as against 40 mm Hg in men (15). Base deficit is 1 mmol/litre for women and 0 for men.

During pregnancy this sex difference is accentuated and the women hyperventilate with a mean Pco_2 of 33 mm Hg (69, 81) and have a compensatory Base deficit of 3—4 mmol/litre.

Ventilation is increased at high altitudes in order to maintain an adequate oxygen supply. Therefore Pco_2 is lower, the higher the altitude. One of the several difficulties in adapting to high altitudes is that again the compensation comes via the kidneys and evolves gradually only over several days. Additional HCO^-_3 is lost via the urine and the relation HCO^-_3/Pco_2 is normalized, i.e. brought towards 0.6.

Hyperventilation begins immediately upon arrival in the mountains, Pco_2 falls and pH increases. Only after a few days is the kidney compensation sufficient to reduce pH appreciably. Meanwhile subjects are likely to be tired and to feel sick. If the higher altitude is reached by walking, however, there is time for normal adaptation. Today when everyone flies, the tourist is faced with more mountain sickness than in earlier days with slower transportation.

There is little knowledge available about the speed of the development of metabolic compensation and there are no valid compensation lines corresponding to Fig. 21. Winter, Dell and Engel (18), realizing that pH often is restored to 7.40, draw a compensation line along the pH 7.40 line in a Pco_2/BE diagram.

In respiratory physiology, for instance in connection with the Olympic games in Mexico City, which lies at an altitude of about 2,000 meters, this is an important problem. In countries where a sizeable part of the population live at high altitudes as in the Andes, this becomes also a social health problem. Other compensatory mechanisms are present such as polycythaemia, but these lie beyond the scope of this presentation.

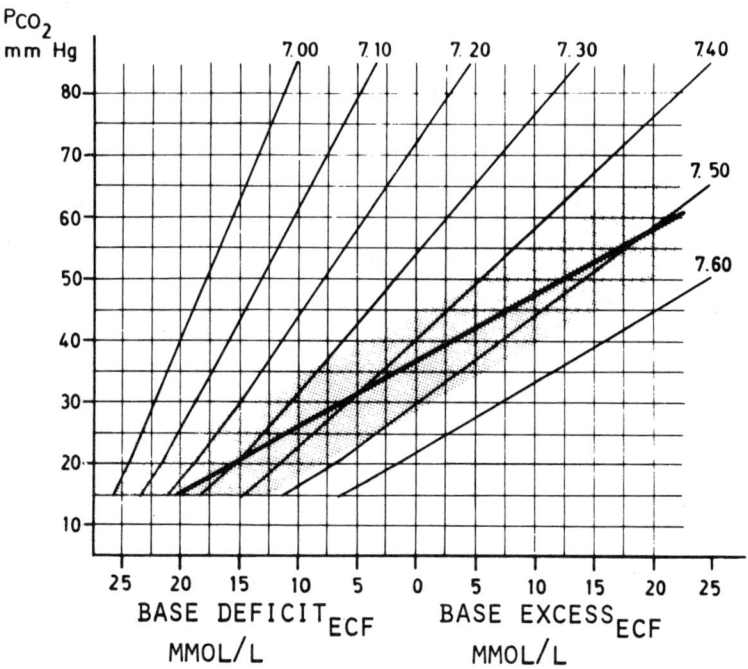

Fig. 23. Respiratory compensation for metabolic acid-base disturbances.
There is a respiratory compensation for metabolic acidosis and alkalosis. Uncomplicated cases will show P_{CO_2} and consequently also pH values, according to the thick oblique line with its 95 percent confidence interval (shaded area).

Chapter 13

Respiratory compensation for metabolic acidosis and alkalosis

It has been noted that the metabolic compensation for respiratory acidosis in uncomplicated cases evolves along established patterns and that the degree of compensation may be predicted from figures such as Fig. 21. In the same way, the respiratory compensation for metabolic acid-base disturbances may be foreseen and a single compensation line may be drawn as in Fig. 23 and is applicable to both primary metabolic acidosis and alkalosis (92). The range is somewhat wider than that for the metabolic compensation for respiratory acidosis.

Clinical examples of respiratory compensation for metabolic acidosis will be given in Chapter 14 on diabetic ketosis and in Chapter 15 on kidney insufficiency.

It is particularly important to recognize the respiratory compensation for metabolic alkalosis, as this leads to a Pco_2 increase. The inexperienced physician might be tempted to interpret such a Pco_2 elevation as a sign of respiratory insufficiency. Another reason for erroneous interpretation could be that now and then, although we really do not know why, there is no respiratory compensation for metabolic alkalosis. Therefore some researchers are of the opinion that no such compensation exists. A clinical example of such a compensation is discussed on page 104 dealing with polyric stenosis and vomiting.

The respiratory compensation for a metabolic acid-base disturbance begins more quickly than a corresponding metabolic compensation, but still needs some time. It usually takes a few hours and it may take as much as 24 hours before the hyperventilation has reached its maximum.

Once the primary metabolic acidosis begins to disappear, it is often noted that the respiratory compensation is still maintained. As a consequence the patient, after his metabolic acidosis with low pH, may easily end up with a high pH value. As discussed in connection with the treatment of diabetic acidosis and lactic acidosis, an iatrogenic metabolic alkalosis is common under these circumstances, also a modest respiratory alkalosis. This hyperventilation, maintained after the primary pH stimulus has disappeared, is probably caused by the slow diffusion of HCO_3^- between the extracellular fluid and the cerebrospinal fluid. If the HCO_3^- concentration in the cerebrospinal fluid remains low, its P_{CO_2} cannot go up without an immediate drop in pH, which would stimulate ventilation. P_{CO_2} therefore increases only pari passu with the bicarbonate concentration in the cerebrospinal fluid.

These changes are usually described with illustrations of the Henderson-Hasselbalch equation, but knowing that the alignment nomogram is nothing but a graphic representation of this equation, we can follow these changes in Fig. 18. Draw a line connecting P_{CO_2} 20 mm Hg and pH 7.20. HCO_3^- is then read as 7.5 mmol/litre. Should P_{CO_2} now be normalized, i.e. 40 mm Hg, but the HCO_3^- level remain the same, pH would fall to 6.9, as seen by connecting P_{CO_2} 40 mm Hg with HCO_3^- 7.5 mmol/litre. Once the bicarbonate concentration increases, P_{CO_2} may increase without any drop in pH.

CHAPTER 14

Diabetic acidosis

Diabetic ketosis is not only the classical example of metabolic acidosis, it is also the classical example of hyperventilation, as the deep breathing of Kussmaul is one of the prominent signs of metabolic acidosis in diabetes. Furthermore, in diabetic ketosis there are profound electrolyte changes and for these reasons this clinical entity is more useful than most for demonstration of electrolyte and acid-base balance disturbances.

Once appreciable changes occur in the electrolytes, the water balance between the different fluid compartments of the body will be influenced as there is always an intimate relation between the amount of sodium and the fluid volume. Moreover, the osmotic effect of the increased glucose concentration profoundly affects the water distribution of the body.

These various mechanisms will be illustrated in discussion of the following case: The patient, who was a 49-year-old woman with diabetes of 15 years' duration, had been admitted several times for diabetic coma or praecoma. Her acid-base and electrolyte values and the changes in these during treatment are shown in Table 4, page 78.

On acidosis

On admittance this patient had a very low pH. (Values below 6.85 are seldom seen.) The cause of the high hydrogen ion concentration (123 nanomol/litre) is of course the excessive degree of metabolic acidosis with a BD_{ECF} of 27 mmol/litre. The patient has partly normalized her pH by hyperventilating, but by extrapolating the compensation line in Fig. 23, it may be seen that Pco_2 should really have been expected to be even lower. The relatively high Pco_2 value indicates that either the compensation was not properly established or the patient was too tired to manage the intense hyperventilation. Without the compensation she would undoubtedly have died because, connecting Pco_2 40 mm Hg and BD_{ECF} 27 mmol/litre in Fig. 18, we read pH as 6.73.

On Day 2 most of the metabolic acidosis had already disappeared. Base deficit$_{ECF}$ was only 10 mmol/litre, which is more or less what an outpatient

Table 4

Electrolyte and acid-base values in a patient with diabetic coma.

	Day 1	Day 2	Day 3	Day 4
Na_p^+ mmol/litre	128		154	140
Cl_p^- mmol/litre	86		91	91
K_p^+ mmol/litre	7.0		2.0	4.0
pH	6.91	7.29	7.54	7.47
H^+ nanomol/litre	123	51	27	33
Pco_2 mm Hg	20	32	39	39
kPa	2.7	4.4	5.2	5.2
BD_{ECF} mmol/litre	27	10	BE 10	BE 4
$Bicarb_p^-$ mmol/litre	3.8	14.5	32.4	28.0
Total CO_2 mmol/litre	4.4	15.4	33.2	29.0

may (but should not) have when, upon undergoing a regular check-up examination, she presents with signs of ketones in the urine, due to a cold.

On Day 3 there was no metabolic acidosis; by contrast the patient had metabolic alkalosis. This is more or less the rule after treatment of serious cases of metabolic acidosis and it is one of the factors which add to the life-threatening character of hypokalaemia, which will be discussed later. The metabolic alkalosis is due mainly to treatment with sodium bicarbonate.

During the first 24 hours the patient was given 240 mmol $NaHCO_3$ intravenously and 180 mmol orally in the form of tablets, i.e. in all a total of 420 mmol.

Mellengaard and Astrup (29) gave the following formula for calculating the amount of sodium bicarbonate needed in order to correct completely a metabolic acidosis:

$$NaHCO_3 \text{ mmol} = \text{body weight (kgs)} \times 0.3 \times \text{Base deficit}$$

The factor 0.3, although experimentally established, roughly corresponds to the total body water and, at least initially, sodium bicarbonate stays mainly in the extracellular fluid. Therefore a factor of 0.15 is more suitable and it is generally recommended that only half of the dose calculated according to Mellengaard and Astrup (29) should be given before the acid-base balance is checked again. Our patient, with a weight of 60 kg, according to the formula would have needed 500 mmol $NaHCO_3$ for a total normalization of her Base deficit if clinical medicine could be reduced to a single and simple formula. The Base deficit of the patient is, of course, influenced by several factors besides the $NaHCO_3$ treatment. Of paramount importance is the rate at which the disease, i.e. the production of 3-hydroxy-butyric acid (3-HB) and acetoacetic acid (AcAc), progresses and the rate at which these acids are metabolized. If the patient is not insulin-resistant, the concentration of 3-HB and AcAc usually begins to decrease within one hour after the insulin injection, but before that the level of free fatty acids and glycerol is reduced, indicating that the abnormal rate of lipid mobilization is reduced (98).

The metabolic acidosis is manifested by an increased plasma level of 3-HB and AcAc. Simultaneously HCO_3^- disappears because the H^+, also added, reacts with HCO_3^- whereby H_2O and CO_2 are formed. The latter, of course is ventilated away. The two reactions on page 42 are both driven to the right by the addition of H^+. The increase in the residual anions will, to express it differently, reduce Buffer base and Base excess, as illustrated in Fig. 24 A.

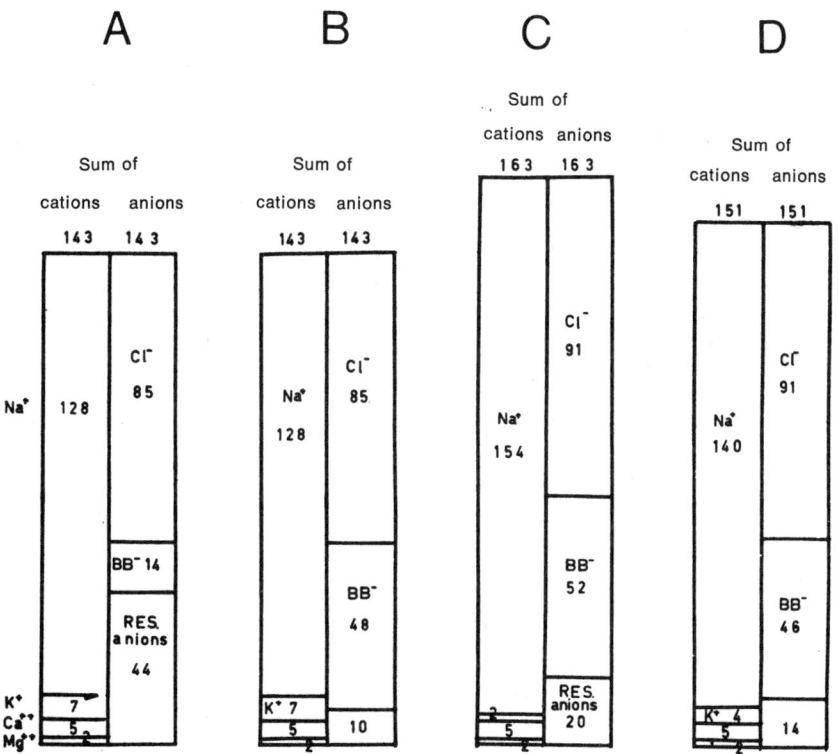

Fig. 24. Gamblegram illustrating the electrolyte changes in a patient with diabetic ketosis
A   at admission
B   hypothetically after insulin administration when the ketoacids are normalized
C   48 hours after admission
D   72 hours after admission

On admittance to hospital, the patient had both low Na_p^+ and Cl_p^- levels, whereas K_p^+ was elevated. If we were now to assume that the treatment consisted only of insulin, which would metabolize the abnormal ketoacids, the pathological increase in residual anions would dissapear. This means a decrease from 44 to 10 mmol/litre, i.e. a ΔBB and ΔBD of 34 mmol/litre. Fig. 24 B shows that in such a case Buffer base would become 48 mmol/litre, which is 6 mmol/litre more than the normal value of 42 mmol/litre. Purely because of the insulin action this patient would then have ended up with a metabolic alkalosis instead of her initial metabolic acidosis. If we calculate the expected effect of the 420 mmol sodium bicarbonate which the patient was given, this would have reduced her Base deficit by 23 mmol/litre.

When sodium bicarbonate is administered there is an increase in the total amount of sodium and water in the extracellular fluid compartment. As a consequence, Cl_p^- is reduced, as schematically indicated in Fig. 25. However, the disease as such is not influenced by this.

When treating a patient with both insulin and sodium bicarbonate, we must realize that both factors increase Base excess.

The variations in the Na_p^+ level shown in Table 4 are due to the fact that the fluid volume changes have been larger than the variations in the total sodium content in the extracellular fluid compartment.

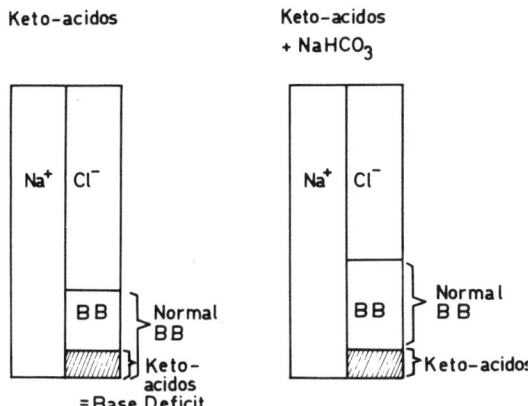

Fig. 25. Schematic representation of the electrolytes in a case of diabetic ketosis before and after sodium bicarbonate administration. It will be seen that when the concentration of residual ions is increased the simple formula $BB_p^- \sim Na_p^+ - Cl_p^-$ is no longer valid. The figure to the right shows that after $NaHCO_3$ treatment Cl_p^- falls and that the primary abnormality, i.e. increase in 3-HB and AcAc, remains although BB is normal.

As a result of the insulin and the sodium bicarbonate treatment, the patient had, on the third day, a Base excess of 10 mmol/litre. She was given NaCl intravenously. As this contains 154 mmol/litre Na^+ and as much Cl^-, she obtained, in relation to the plasma level, more chloride than sodium. The following day both the electrolytes and the acid-base balance were normal.

A symptomatic correction of pH should be considered if pH is less than 7.15 and always be made if pH is lower than 7.10 ($H^+ < 80$ nanomol/litre). Without such a pH correction neither the circulation nor the cellular metabolism will function properly and it is possible that the insulin effect is also better if pH is closer to normal. Therefore, in diabetic coma the first fluid to be administrated is 0.5 litre of the isotonic $NaHCO_3$ solution (containing 167 mmol $NaHCO_3$/litre). This should be followed by 1 litre isotonic saline. Subsequently $NaHCO_3$ and NaCl should be given alternatively, according to the acid-base and electrolyte analyses.

If the patients are not in coma and have only a moderate degree of ketosis, there is usually no need to give sodium bicarbonate intravenously. After the insulin injection, it suffices to give NaCl parenterally and sodium bicarbonate orally. Remember that the insulin takes care of the residual anion increase.

The metabolic alkalosis so often seen after the successful treatment of diabetic coma and mainly iatrogenic in origin, is no clinical problem except for the fact that it adds to the risk of hypopotasaemia.

On potassium

The plasma potassium level is to a large extent a function of the hydrogen ion concentration. When this is high, K_p^+ is high, and vice versa. Expressed in pH units, K_p^+ is high when pH is low, and K_p^+ is low when pH is high. In a series of patients having small Pco_2 changes, Rooth and Fürst (120) found the following relation:

$$K_p^+ = 3.5 - 0.2 \times BE$$

In Table 4 pH stands below K_p^+ and the inverse relation between these two factors in this patient is clearly seen.

There are two main reasons for the K_p^+ fall during the treatment of coma:
1. The pH increase
2. The restoration of potassium depletion of the cells.

It is therefore a mistake to wait with the potassium administration until K_p^+ is lower than normal. Instead, potassium should be added to the infusion as soon as K_p^+ begins to fall from high towards normal values.

The total electrolyte losses during ketosis are considerable. Maxwell and Kleeman (9) give the following mean values:

Na^+ 8 mmol/kg bodyweight
K^+ 6 ,, ,,
Cl^- 5 ,, ,,

Calculated for a man of 70 kg, the mean potassium loss is 420 mmol out of a total store of 3000—3500 mmol. It follows that about 10—15 % of the total amount of potassium in the body will be lost during the course of a few days of diabetic praecoma and coma. All this K^+ comes from the cells and must be restored to the cells. It is obvious that the measurement of K_p^+ gives no information about these losses and very limited guidance as to the need for potassium.

There are several factors adding to the potassium loss. One of the most important is the general tissue breakdown which occurs in diabetic coma, and which is best measured as nitrogen loss. As the intracellular potassium concentration is some 20 times higher than the plasma level, the cellular destruction leads to a flow of potassium into the extracellular fluid and a subsequent elimination via the urine. Potassium is also bound to glycogen and released when, as in ketosis, the cellular glycogen stores are depleted. Potassium is also liberated from the cells during dehydration. Finally there is often an appreciable potassium loss from vomiting; some 10 mmol is lost with each litre of vomitus.

As soon as the therapy becomes effective, the direction of the potassium movement is reversed and the need for supplementary potassium is great, as the cells cannot obtain much from the extracellular fluid.

In clinical medicine it is not possible to measure the total potassium losses; therefore we must content ourselves with some simple rules for estimating these. Such rules suffice only to give an indication as to planning the treatment. Further guidance must be obtained from repeated measurements. Table 6, page 102, gives some information about the electrolyte losses in different body fluids. Another feasible method is:

1 to estimate the total water loss.
2 to estimate that half of the water loss is derived from the intracellular- and half from the extracellular fluid.
3 to gauge the total electrolyte losses from the composition of the extra- and intracellular fluids (Fig. 11 and Table 1, respectively).

If we assume that a patient weighs 60 kg and has had a total water loss (judged from the total weight loss) of 6 kg, this corresponds to 10 % of total body weight, or 20 % of total body water.

The extracellular fluid loss of 3 litres is readily compensated for by giving an isotonic sodium-acetate solution (failing this, Saline or Ringer solution will do).

Thus the intracellular fluid loss of 3 litres would correspond to:

$3 \times 150 = 450$ mmol potassium
$3 \times 10 = 30$ mmol sodium
$3 \times 15 = 45$ mmol magnesium

and of the anions mainly phosphate

$3 \times 150 = 450$ mmol phosphate.

Although this method is not too correct in diabetic coma because the water losses are proportionally larger than the electrolyte losses, as will be discussed below, it follows that the patients need large amounts of potassium. The time of onset and the dosage of potassium substitution is therefore a cardinal point in the treatment of diabetic ketosis.

The old rule that potassium should be given when there is hypopotassaemia is no longer valid. Hypopotassaemia should not be treated; it should be prevented. The safest way to achieve this is to initiate potassium substitution as soon as a high K_p^+ begins to fall. In those cases where K_p^+ is normal or low already at admission, the potassium substitution must be given together with the first litre of fluids.

Plasma potassium may fall very rapidly after the onset of insulin and sodium bicarbonate therapy in the diabetic coma patient; there are patients on record who have died on the first day of hypopotassaemia. Alberti et al (87) therefore recommend potassium administration at the same time as the first insulin injection. They gave a mean value of 70 mmol potassium per day and as a result K_p^+ fell in those patients initially having high values and rose in those patients initially having low K_p^+ levels.

On water

The disturbance in the water balance in diabetic coma is considerable. Remember that Na_p^+ really says more about the fluid volume in the extracellular compartment than about the amount of sodium itself. On admission, to hospital, our patient in Table 4 had a Na_p^+ of 128 and Cl_p^- of 85 mmol/litre, i.e. hyponatraemia and hypochloraemia were present at the same time.

This may be due to one or a combination of the following causes:

1) There is a net loss of sodium and chloride from the extracellular fluid.
2) There is a dilution of the extracellular fluid. This in turn may be due to

a. a total increase of body water
b. a shift of intracellular water to the extracellular compartment
c. a combination of factors a and b.

A dilution of the extracellular fluid will of course be apparent mainly in the dominant ions. A 10 % change in the calcium or magnesium level would perhaps still be within the normal range and the plasma potassium concentration, as stated, is much influenced by other factors. Serial measurements of plasma protein, haematocrit, haemoglobin or red cell levels may help materially in indicating haemodilution or haemoconcentration, but Na_p^+ measurements are usually equally satisfactory in this respect.

In diabetic ketosis there are large losses of sodium, as shown, but there is also a dilution of the extracellular fluid in spite of the obvious fact that these patients have lost large amounts of fluid and consequently have a pronounced loss of total body water. The total loss of water in diabetic ketosis may well be as high as 10 % of the total body weight, or roughly 20 % of the total body water.

In this context it may be useful to recall the figures for total body water. In females of 40 to 60 years it is 45 % and in males 55 % of the body weight. In younger people the figures are higher. Fat contains only 5—10 % water; the fatter a person, the less is his relative water content.

The combination of dilution of the extracellular fluid and dehydration of the intracellular fluid occurs only in diabetic ketosis and in hyperosmolar, non-ketotic diabetic coma. Only in the final stages of these diseases do patients lose so much water that the plasma concentration also increases.

High Na_p^+ or Cl_p^- in a case of diabetic ketosis is a sign of extreme dehydration and there is an imminent risk for circulatory collapse.

The mechanism of the hypervolumaemia usually seen is interesting. The normal osmolarity of the extracellular fluid is 285 mOsm/litre, and it is the same in the intracellular fluid. If the osmolarity of the extracellular fluid compartment increases, water is drawn from the cells until iso-osmolarity is restored. In diabetic ketosis and hyperosmolar diabetic non-ketotic coma, the blood glucose level is markedly increased and, as a result, the osmolarity is also raised.

The molecular weight of glucose is 180. Consequently 180 g of glucose in 1 litre of water gives 1 mol/litre and 1 Osm/litre. When, as we should see within a few years, glucose also is given in mmol/litre, it will be much easier to realize at once the osmotic effect of glucose. As 90 mg glucose/100 ml of blood gives 5 mOsm/litre, 180 mg/100 ml of blood gives 10 mOsm/litre and 810 mg/100 ml gives 45 mOsm/litre. The increase in the glucose level is only in the extracellular fluid as the glucose cannot enter the cells without insulin. Thus an osmotic gradient is built up, drawing water from the intra-

into the extracellular fluid compartment. As a consequence, sodium and chloride are diluted, as are all the other dissolved particles in the extracellular fluid. Na_p^+ therefore initially is lower, the higher the glucose concentration. In the final stages of coma the water loss is such that both the intra- and the extracellular fluid compartments are dehydrated, whereupon Na_p^+ rises.

In diabetic coma the osmolarity increases from 285 to about 320 mOsm/litre and may become as high as 370 mOsm/litre. As soon as the insulin action takes place, glucose will again penetrate into the cells and the osmotic gradient will disappear. This causes a water flow back into the cells.

To some extent the relation between water and electrolytes is illustrated in the Gamblegrams in Fig. 24, page 79. In A the height of the column of the electrolytes was rather low, a sign of dilution. In C the column was high, indicating some plasma dehydration. In D the sum of the electrolyte concentrations was normal and the patient was normal as regards water, osmoles, electrolytes and acid-base parameters. If the sum of the cations markedly differs from the usual values of 153 meq/litre, a fluid problem should always be suspected. However, if there is a concomitant change in water and sodium in the same relation as in plasma, the height of the Gamble column will not change, despite ample infusions, as when isotonic sodium chloride is administered.

Diabetic patients have their main water loss via urine, but larger-than-normal amounts of water are also lost via the lungs, and vomiting is common in diabetic praecoma. The increased urinary volume is caused by several osmotic factors. The glucose elimination increases the urinary osmolarity, and at the same time the sodium reabsorption is reduced so that there is a marked loss of sodium. Also, large amounts of potassium are eliminated as a result of the cellular breakdown and dehydration. Among the anions, the most important are 3-hydroxybutyrate and acetoacetate. Together they may give a concentration of 300 mmol/litre or more. In summary it may be said that glucose, sodium, and potassium chloride, ammonium butyrate, and ammonium acetoacetate account for the increased osmolarity of the urine. In diabetic ketosis there is also an increased NH_4^+ production in the kidneys. This then is an example of a metabolic compensation for a metabolic acidosis.

The normal water loss via the lungs is about 400 ml/24 hours. It is proportional to the alveolar ventilation and the temperature of the patient. In diabetic ketosis, the water loss via respiration may be as high as 1.500 ml/24 hours. If all the fluid losses are added, also all the electrolyte losses, it will be seen that the patient has lost relatively more water than electrolytes. Another way to express this is to say that the patient has a loss corresponding to a hypotonic solution.

The fluid loss via the lungs contains no electrolytes, vomitus is hypotonic, and the urine has lower sodium and chloride concentrations than plasma. This must be remembered when the patient is treated. Even if the patient has lost large amounts of electrolytes, he has lost relatively more water. As a rule this causes no therapeutic problem. It suffices to give isotonic electrolyte solutions intravenously and water orally. If the patient cannot drink adequately the same result is obtained by giving glucose (or other sugar solutions) intravenously.

Hyperosmolar nonketotic diabetic coma

The salient features in diabetic coma are metabolic acidosis and ketosis. In 1957 another type of diabetic coma was described in which no acetonuria was present, and no metabolic acidosis. The blood glucose values were very high as was the osmolarity. The patients are usually elderly with a mean age of about 60 years. The only young man I found a reference to was one of the cases reported by Seften et al (99), but this is one of many in which the diagnosis seems inappropriate as the patient had a bicarbonate level of about 8.0 mmol/litre.

As in any diabetic coma, the glucose increase is limited to the extracellular fluid and the cells are "starved" for glucose. The extracellular fluid osmolarity increases because of the glucose increase and water is drawn from the cells in order to reestablish the iso-osmolarity. The osmotic diuresis is prominent, as well as the total dehydration. The coma is caused either by the osmolarity or by the hypovolaemia, or both. The osmolarity in these cases increases from 285 to 320 mOsm/litre and figures as high as 440 mOsm/litre have been published.

As a rule patients with hyperosmolar non-ketotic diabetic coma should have no acid-base disturbances, but there are probably borderline cases which could as readily be classified as ordinary diabetic coma. A few of the cases on record have had metabolic alkalosis (BE 20 mmol/litre or higher), but in these cases the patients were given thiazides (see page 107). The electrolyte concentrations of these patients will of course depend upon the dilution or concentration of plasma, just as in any other diabetic coma. Therefore in the final stages both Na_p^+ and Cl_p^- are elevated, indicating severe hypovolaemia.

The treatment consists of insulin and large amounts of fluids. Insulin will admit glucose into the cells and take away the osmotic gradient between the extra- and intracellular fluid compartments. As both glucose and water reenter the cells, it is most important to give so much fluid intravenously that the inital hypovolaemia is not aggravated by the insulin therapy.

Usually these patients need some 8 litres during the first 12—24 hours, but there is no unanimous opinion about how this should be done. Some advocate the administration of isotonic NaCl with the addition of KCl, others the infusion of 5.5 % glucose solutions. The latter may seem paradoxal and should not be used before it is known that the blood glucose begins to decrease as a result of the insulin effect and when the patient has no hypovolumia. Saline stays virtually only in the extracellular fluid, whereas glucose solution is distributed evenly over the total body water.

As patients lose considerable amounts of electrolytes during polyuria, it would seem that a judicious use of both Saline and glucose solutions would be advantageous. The order of administration would be 1) insulin 2) Saline 3) 5 % glucose solutions and, subsequently, electrolytes or glucose, depending upon the plasma electrolyte composition.

Lactic acidosis

Huckabee (92) was the first to describe a syndrome characterized by severe metabolic acidosis with coma and usually resulting in death within a few days. He showed that the metabolic acidosis was caused by an increase in the lactate leading to a decrease in Buffer base. H^+ from the lactic acid had united with HCO_3^- forming CO_2 and H_2O, and the CO_2 thus formed disappeared with the ventilation. Some H^+ reacted with the plasma proteins, i.e. both reactions given in page 42 were driven to the right, just as in diabetic ketosis.

We usually do not know what initiates a lactic acidosis. Possibly it is a terminal sign much more frequently than we realize. The particular clinical entity which is called lactic acidosis evolves usually during the course of several days and the patient gradually becomes more and more comatose, showing only a modest respiratory compensation for the metabolic acidosis, and consequently a low pH. It is not known what causes the lactoacidosis. There are two theories, however: one is that the circulation is insufficient and therefore also the oxygen supply to the tissues: a manifestation of tissue hypoxia; the other hypothesis is that the metabolism of lactic acid is reduced in the liver. We do know that patients with kidney insufficiency are more disposed to lactic acidosis than others, particularly if they have been given phenformin.

As a rule lactic acidosis is irreversible, but sometimes it is possible, by means of large amounts of sodium bicarbonate, to raise the pH so much that the vicious circle is reversed. Nevertheless, there may be considerable difficulties in the management of the patient, as illustrated in Fig. 26. In Robin's article (96) the metabolic component is given as HCO_3^-, but a free translation into Base excess is made here.

At the onset of the treatment the patient had a Base deficit of 25 mmol/litre and was given a total of 800 mmol NaHCO$_3$ during two hours. The metabolic acidosis still remained at the same level. The implication of this is that the pathological process progressed at the same rate that the sodium bicarbonate was injected. If the disease process had ceased, the 800 mmol NaHCO$_3$ would have more than normalized Base deficit.

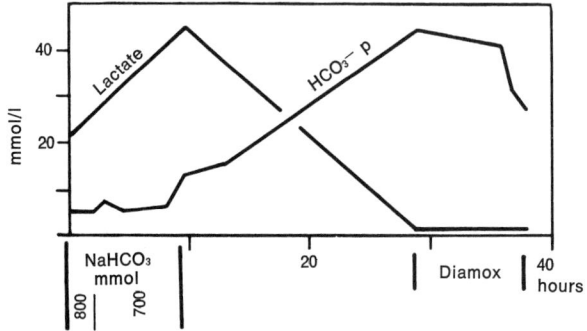

Fig. 26. Lactate and bicarbonate changes in the course of treatment of a patient with lactic acidosis.
Modified after E. Robin (96).

Fig. 26 shows how, although the HCO$_3^-$ remained constant, the lactate concentration rose, as a sign of the progress of the disease. The rate of this progress may be calculated from the sodium bicarbonate dosage to be 400 mmol of lactic acid per hour. Calculated per 24 hours, this would correspond to 9,000 mmol or about half of the total amount of lactic acid normally produced and metabolized. Whether in this case the production was increased as in hypoxia, or only the metabolism was reduced, is not known.

Once the inital 800 mmol NaHCO$_3$ were given it was enough to give some 70 mmol/hour to keep Base deficit constant and to maintain pH at the same level. When HCO$_3^-$, and probably also Base deficit, was about 12 mmol/litre, the sodium bicarbonate therapy was withheld.

At about that point it was noted that the lactate level began to be reduced, where upon there was a steady fall in the lactate level and increase in the HCO$_3^-$ concentration. After some 20 hours BD = BE = 0. However, note that the lactate level at that time was still 20 mmol/litre. We see here yet another example of how metabolic acidosis is *compensated* for by sodium bicarbonate. Compare this with the situation in diabetic ketosis and kidney insufficiency.

When subsequently the lactate level fell consistently to become normal 10 hours later, Base excess gradually increased to about 20 mmol/litre 30 hours after the onset of the administration of sodium bicarbonate. At this time pH was probably well over 7.50. The patient was then given acetazolamide (Diamox®) which causes a loss of $NaHCO_3$ in the urine (see chapter 17) and 10 hours later both lactate concentration and BE were normal.

That the patient, if he survived, should end up with a metabolic alkalosis after the administration of 1,500 mmol of sodium bicarbonate is no surprise. If anything, the metabolic alkalosis could have become more pronounced. In our calculations as to the effect of the sodium bicarbonate we would have needed the body weight of the patient, but this is not given.

One simple rule says that the administration of 1 litre of isotonic $NaHCO_3$ (167 mmol) increases BE 10 mmol/litre, and according to the formula given earlier:

$$\text{the } NaHCO_3 \text{ need} = BD \times 0.3 \times kgs$$

If the weight of the patient was 70 kg, 1,500 mmol $NaHCO_3$ would increase BE 90 mmol/litre according to the first rule and 70 mmol/litre according to the formula. As the total BE increase was only about 45 mmol/litre, the weight of the patient was probably 90—100 kg.

CHAPTER 15

Renal insufficiency

Whenever the fluid, electrolyte, or acid-base balance of a patient is considered, his kidney function must also be evaluated. On page 66 a description is given of the process by which the metabolic compensation for respiratory insufficiency is achieved by the kidneys, which produced added amounts of NH_4^+. In the present context the kidneys have the following three functions:

1. Regulation of the amount of body water
2. Regulation of the amount of body electrolytes (mainly sodium and potassium)
3. Regulation of the metabolic acid-base balance of the body.

In renal disease one or several of these functions may be more or less reduced. Even if, in the individual case, it may be difficult to obtain exact information about the scope of the particular kidney deficiency, it is worthwhile to consider these three factors and to estimate if function is reduced or not. Furthermore, such an evaluation must be repeated from time to time as the disease progresses or the therapy becomes effective. A patient with renal insufficiency may on one day have oedema, hypertension, and headache, as a sign of salt and fluid retention, and a few days later may have fluid and salt depletion.

The kidney contributes to the regulation of the amount of body water by producing according to needs:

    a) concentrated urine
    b) diluted urine

The words "concentrated" and "diluted" are used here in relation to plasma, the specific weight of which is 1.010, or better expressed, the osmolarity of which is 285 mOsm/litre. Urine more concentrated than plasma has a higher osmolarity than 285 mOsm/litre, and vice versa. Healthy kidneys may concentrate the urine to a specific weight of 1.030 or an osmolarity of about 1.000 mOsm/litre and dilute to about 50 mOsm/litre. Undoubtedly the osmo-regulating function of the kidney would be better understood if the old expression "specific weight", were abandoned and we spoke only about the "osmolarity" of the urine.

Diseases which affect mainly the renal marrow tend to reduce the ability of the kidney to concentrate the urine. Such diseases are chronic pyelonephritis, analgesia nephropathy, nephrocalcinosis, and hereditary nephronophthisis. If the renal filtration is severely reduced in glomerulonephritis these cases, too, will have isotonic urine.

The kidneys eliminate metabolic waste products such as salt or other solutes. Ordinarily these add up to some 600 mOsm/24 hours. If the concentration power is reduced, these 600 mOsm must be eliminated in a larger volume of urine. When the kidneys have lost all their ability to concentrate and the urine therefore has the same tonicity as plasma, it takes 2.3 litre of urine to dissolve 600 mOsm. This, then, is the minimum urine volume such patients must have. As they also have reduced ability to eliminate water, all the 24 hours must be used for drinking and urinating. This means in clinical practice that when they awaken during the night to urinate they must also drink a corresponding amount. Both the patients and the nursing staff must be carefully instructed, otherwise the patients will become uraemic. For these patients, water prescription is just as important as any drug prescription.

The pratical difficulty often lies in finding a suitable middleway between, on the one hand, too little water intake, too low urine volume, and uraemia, and, on the other hand, too much water intake, water retention, increasing blood pressure, oedema, and headache. The patients should therefore be weighed every day under standard conditions and the weight should remain constant. The best results are usually achieved if the patients are instructed to draw graphs themselves of their daily body weight, water intake, and urine volume. After a short time they manage these things very well and the responsibility of the nursing staff is reduced. Also, the patient manages much better at home than otherwise.

If the kidneys cannot eliminate the daily 600 mOsm, the waste products must be reduced. These derive mainly from the metabolism of protein, and are to a large extent nitrous compounds, sulphate, and phosphate. Both carbohydrates and fat end up as CO_2 and water, and produce no urinary osmoles. The relation between osmoles, calories, and water production of these different substances is given in Table 5.

Ordinarily both exogenous and endogenous proteins are metabolized. Therefore the protein waste products cannot be abolished by eliminating protein ingestion because the endogenous protein breakdown then increases. But a marked reduction of protein intake coupled with a diet rich in carbohydrates and fats and containing, supplementarily, the essential aminoacids, may materially reduce the amount of waste products.

Remember that 600 mOsm of waste solutes are produced by healthy persons. After tissue destruction for any reason: catabolism, surgical trauma, infection, or inflammation, treatment with steroids or cytostatic agents, or following haematomata, the protein breakdown is markedly increased. Patients who have earlier had adequate renal function may develope a life-threatening uraemia without any further reduction in the filtration of the kidney. In these cases it is better to measure plasma urea than creatinine as a sign of uraemia. Remember in all such cases to discuss the possibility of haemodialysis early, before it is too late.

Table 5

Production of calories, water, and osmoles by combustion of 100 g

	calories	water ml	mOsm
Carbohydrate	400	50	–
Fat	900	100	–
Protein	400	300	500
(given as 400 g of meat)			

In acute tubular necrosis caused by trauma, poisoning, or blood transfusion with incompatible blood groups, there is, as a rule, an anuric phase at first, followed by polyuria, when the patient may pass four to five litres of urine a day. As this urine contains salts, the patients lose solutes, particularly sodium. After having had high blood pressure and low urine volume the patient becomes dehydrated because of large urine volumes and has low blood pressure, initially only when measured standing. In spite of the extent of diuresis, it is not infrequently noted that after an initial decrease in urea level this again begins to rise at a time when it would be expected that the risk of uraemia was already passed. In this situation repeated analyses of plasma sodium must be made and preferably also of the total urinary sodium elimination. The losses must be adequately compensated for. A fall in Na_p^+ is usually followed by azotaemia. The reason is possibly that the body must maintain normal osmolarity. This relatively frequent cause of azotaemia is treated by raising Na_p^+ to a normal level. It is worthwhile to remember the old French medical adage "azotémie par manque de sel" in any case of uraemia, and to measure Na_p^+. (See Fig. 27.)

The ability to produce diluted urine, i.e. with lower osmolarity than plasma, usually follows normally after intake of great amounts of fluids, such as during beerdrinking contests. Patients with diabetes insipidus can produce only hypotonic urine. This is one of the salient features of that syndrome.

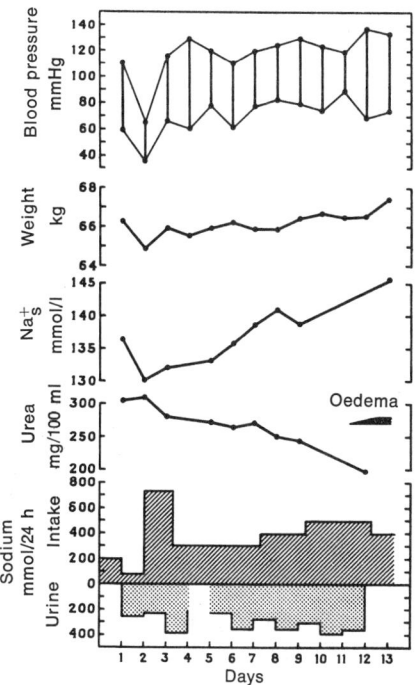

Fig. 27. Azotaemia by hyponatriaemia.
The urea concentration was normalized when so much NaCl was given that Na_p^+ became normal. Note the fall in the blood pressure (measured standing) when Na_p^+ was low.
Modified after J. R. Cove-Smith & M. S. Knapp (102).

The primary disturbance is often in the posterior lobe of the hypophysis which cannot produce an adequate amount of antidiuretic hormone (ADH), but, particulary in some hereditary forms, the deficiency is localized to the kidneys, which do not respond to ADH.

In most cases where there is an overall reduction of renal function, such as in hypertensive disease, chronic pyelonephritis, or diabetic nephropathy, the kidney still retains the ability to dilute the urine, but this reaction comes slower than is normally the case. A healthy kidney increases the diureses after a water load within one or two hours. When the kidney function is impaired it may take six to ten hours before an equivalent renal response is seen. Such patients may easily be given too much fluid if they are given drink or infusions too fast.

In the treatment of patients with uraemia it should be remembered that uraemic pulmonary oedema is interstitial and therefore difficult to diagnose by auscultation, and that psychic changes, somnolence, or convulsions are often caused by cerebral oedema. Uraemic toxicity as such, or together with the disease which causes the renal damage, often leads to extracapillary transudation; it is therefore frequently advisable to be somewhat restrictive in salt and fluid substitution.

As stated above, the lungs and the kidneys together regulate the acid-base balance of the body. The lungs regulate the respiratory component, the kidneys, the metabolic component. The lungs function rapidly in this respect by increasing or decreasing the CO_2 elimination. The kidneys perform their function slowly by increasing or decreasing the H^+ elimination in two ways:

a) via the phosphates
b) as NH_4^+

The amount of free hydrogen ions eliminated by the urine is so small that it may be dismissed.

Renal disease which affects the acid-base balance always leads to a metabolic acidosis, although various mechanisms may be involved.

In renal insufficiency the elimination of various waste products is reduced. Not only do creatinine and urea increase, but also plasma, phosphate and sulphate. This is schematically illustrated in Fig. 28.

Base deficit in Fig. 28 A is 42—32 = 10 mmol/litre. According to the compensation line in Fig. 23, P_{CO_2} at that BD level should be about 27 mm Hg. We can then calculate pH from any of the nomograms and find that the patient probably has the following acid-base status:

pH 7.35 (45 nanomol/litre)
P_{CO_2} 27 mm Hg (3.6 kPa)
BD_{ECF} 10 mmol/litre
HCO_3^- 14 mmol/litre

In textbooks on kidney disease it is usually stated that there is no cause to treat the metabolic acidosis as such before the bicarbonate concentration is below 15 mmol/litre. As indicated by the above example, we can also say that no special therapy for the acidosis is needed until Base deficit is more than 10 mmol/litre.

If the respiratory compensation is normal, i.e. follows the compensation line in Fig. 23, page 74, the pH value is only slightly reduced and no therapy is indicated because of the pH level itself. However, with increasing metabolic acidosis the respiratory compensation increases and the patient often experiences dyspnoea when he must maintain sufficient ventilation to keep

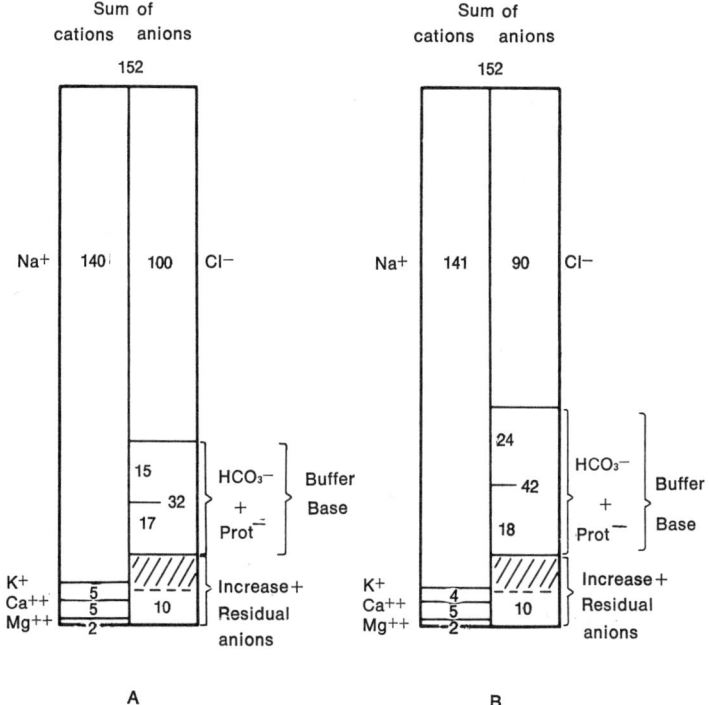

Fig. 28. Gamblegrams showing metabolic acidosis in renal insufficiency and the effect of sodium bicarbonate therapy.
The patient had a Base deficit of 10 mmol/litre and electrolytes according to Fig. A initially and to Fig. B after treatment with so much sodium bicarbonate that Base deficit was 0.

Pco_2 below 30 mm Hg. A simple rule is, therefore, that the metabolic acidosis should be reduced when the patient suffers from dyspnoea.

The simplest way to normalize Base deficit, and secondarily pH and Pco_2, is to give the patient tablets with 1 g of sodium bicarbonate three times daily, or, if necessary, twice that amount. The maintainance dose should only just suffice to keep BD between 0 and 5 mmol/litre. Note, that as always after the administration of sodium bicarbonate, Base deficit, Pco_2 and pH may be normal, but the primary acid-base disturbance is still present in the residual anions and the patient happens to have BE 0 because he simultaneusly has a metabolic acidosis of 10 mmol/litre and an iatrogenic metabolic alkalosis of 10 mmol/litre. Note also that the sodium concentration is scarcely affected

by the therapy, whereas the chloride level is as much reduced as Buffer base (and BE) is raised. His Cl_p^- decrease results from:
1) a dilution of the extracellular fluid volume or
2) an increased urinary elimination of Cl^-.

The administration of sodium bicarbonate leads mainly to an expansion of the extracellular fluid compartment. Therefore the patient must be weighed regularly in order to detect and prevent pathological fluid retention. For this reason sodium bicarbonate should not be given too fast and the dosage should be increased when it is apparent that the patient does not gain weight unduly. Patients with water-and-salt-losing renal disease often need some supplementary sodium administration such as $NaHCO_3$ although they have only small changes in Base deficit.

When treating patients with cardiac arrest symptomatically with sodium bicarbonate, in order to normalize pH, basically the same changes occur as in Fig. 28 A and B. The abnormal concentration of residual anions remains unchanged until the circulation is restored and the lactate is metabolized. The cardiac arrest situation is therefore similar to that seen in diabetic ketosis as described on page 80.

However, there is a difference between cardiac arrest and diabetic ketosis on the one hand and renal insufficiency on the other. In the first group the abnormal residual anions are metabolized, whereas this does not take place in the second group. The end result of a successful treatment of cardiac arrest or diabetic ketosis is, therefore, an iatrogenic metabolic alkalosis, but this does not occur in renal insufficiency. In Fig. 28 A the patient had a metabolic acidosis with respiratory compensation. In Fig. 28 B he had, as stated above, a metabolic acidosis balanced by metabolic alkalosis.

Clinical conditions in which the acid-base regulation of the kidneys is reduced

Normally only small amounts of NH_4^+ are produced. The ammonium generation should be seen as the reserve capacity of the kidney in response to an increase in hydrogen ion load. The maximal NH_4^+ production is achieved only after five to seven days and may then be about 500 mmol/24 hours. It is relatively easy to test the renal response to an acid load. This may be done during the course of one day with the fast ammonium chloride test according to Wrong and Davies (111). Liver function must of course be normal before ammonium chloride is given, as in the liver the reverse reaction takes place from that in the kidney shown on page 66. The NH_3 liberated must then be detoxified by the liver cells.

The ability to produce NH_4^+ is reduced in various kidney diseases. It follows that patients subjected to an acid load are more disposed to metabolic acidosis than are others, and if they also have respiratory insufficiency they cannot compensate according to Fig. 21, and their pH drop, for a given P_{CO_2} increase, will be correspondingly larger.

Phenformin administration also reduces renal ability to produce NH_4^+ (108), which probably is the reason why severe acidosis is comparatively frequent if Phenformin is given to patients with renal disease.

Patients with renal tubular acidosis have more or less normal glomerular function whereas their tubulus function is defective. Urinary pH is, in these cases, relatively fixed between 6.5 and 7.0. The normal Na^+ and H^+ exchange is reduced and both K^+ and HCO_3^- are lost in the urine. Usually the cation Na^+ is reabsorbed from the lumen and balances HCO_3^-. In exchange H^+ is eliminated and coupled to phosphate according to:

$$Na^+ + HPO_4^{2-} + H^+ \rightleftharpoons Na^+ + H_2PO_4^-$$

Patients with renal tubular acidosis seldom have pronounced metabolic acidosis as such and no specific treatment is needed for the acidosis. However, the concomitant hypokalaemia cannot be tolerated, otherwise it may lead to hypokalaemic nephropathy. This is one of the few situations where one finds the combination of hypokalaemia and metabolic acidosis and when it may therefore be better to give potassium citrate than potassium chloride.

Renal tubular acidosis is also one of the symptoms in the Fanconi syndrome which belongs to the clinical entity: inborn errors of metabolism.

Newborn infants, mainly premature infants, may develope metabolic acidosis because of insufficient phosphate ingestion. The kidneys then eliminate less phosphate than normally and less H^+ may be trapped in the above reaction.

By means of pH titration of the urine it is possible to measure the amount of hydrogen ion eliminated as secondary phosphate and the amount bound as NH_4^+. This is, moreover, an important method to gauge the total acid-base balance of the body at the same time that it gives information about the kidney function in this respect. It is therefore somewhat surprising that such titrations are so seldom performed. The analytical procedure is easy and reliable. The only difficulty lies in obtaining the total urine volume for a certain time period. This difficulty is much greater in clinical practice than most people realize, and it is a difficulty which is relevant for most urine determinations.

Hyperkalaemia is a life-threatening complication in patients with renal insufficiency in whom, for one reason or another (see above), increased tissue destruction occurs. The treatment then must be instituted early. It helps to remember this complication and to realize the advantage of the prompt onset of therapy.

Plasma potassium may be reduced by the following procedures:

1 By making the patient alkalotic (or less acidotic) with sodium bicarbonate
2 By the administration of glucose and insulin
3 By ion exchange resins given as enema or orally
4 By peritoneal dialysis
5 By acute haemodialysis

ad 1 The relation between H^+ and K_p^+ has already been described several times.

ad 2 On page 82 it was explained how potassium was liberated when glycogen was mobilized in diabetic coma and how, later, K_p^+ fell when the patients had been given insulin, the glucose began to reenter the cells, and glycogen could be restored.

ad 3 Different drugs are available in different countries. Therefore references must be found in each country's pharmacopoeia.

ad 4 In peritoneal dialysis, potassium is dialyzed over the biological membranes in the abdominal cavity and eliminated by the dialyzing fluid. This treatment can be instituted at any hospital within a short time.

ad 5 The blood is dialysed by an artificial kidney. Only specialized hospital departments possess the necessary facilities and experience. Haemodialysis is more effective than peritoneal dialysis. In the case of a patient living at some distance from the specialized kidney unit, it may be advisable to begin peritoneal dialysis in the local hospital and later to transport the patient to a dialysing renal unit.

A few points particularly important in the evaluation of patients with renal disease are here again brought to mind:

The fluid and electrolyte balances are intimately correlated and cannot be separated one from the other. The sodium concentration therefore often gives better information about the total amount of fluid in the extracellular fluid than about the total amount of sodium. If there is a lack of water, there is usually a lack of sodium, and vice versa. The sodium concentration which we measure as Na_p^+ then indicates only whether the sodium loss is more pronounced than the water loss, or the water loss more pronounced than the sodium loss.

It is therefore common to note low Na_p^+ values in spite of high total sodium amounts in the extracellular fluid, mainly in patients with acute glomerulonephritis or severe hypertension. "Correction" of Na_p^+ by salt administration in such patients may precipitate a hypertensive crisis, pulmonary and cerebral oedema. In hypovolumia depending upon water losses, some patients may show so-called water depletion values with a high Na_p^+ while other patients, particularly if their urinary or gastroenteral salt losses have been conspicuous, may have low or normal Na_p^+ levels.

There is a special system which regulates the osmolarity of the body to a constant level of about 285 mOsm/litre. Roughly this corresponds to $2 \times Na_p^+ = 2 \times 143 = 286$ mmol/litre. Every mmol gives a mOsm. The amount of water is regulated by the antidiuretic hormone of the posterial lobe of the hypophysis and ADH acts on the water reabsorption in the renal tubuli. The amount of salt eliminated in the urine is governed by the aldosterone produced in the adrenals, acting upon the ionic exchange in the tubuli. Thus more ADH is produced and water retained, in order to keep a normal osmolarity if there is too much salt (NaCl) in the body, or, if there is primary water retention, the aldosterone secretion increases and less sodium is eliminated via the urine.

CHAPTER 16

Acid-base disturbances due to electrolyte changes

Chapter 5 dealt with so-called primary acidosis, i.e. when either the normal elimination of H^+ as CO_2 via the lungs or of H^+ via the urine was restricted, or there was an abnormal production of acid, as in diabetic ketosis or lactic acidosis. There is also a group of secondary acid-base disturbances which are due to electrolyte losses (or administration). In such situations there is no change in the residual anion concentration and the simplified equation: $BE_p + 42 \sim Na_p^+ - Cl_p^-$ is valid. It follows directly from this equation that if the body loses more chloride than sodium in relation to the plasma concentration of these substances, there will be a metabolic alkalosis, and, conversely, if the sodium loss is larger than the chloride loss, there will be a metabolic acidosis. The H^+ or OH^- movements to or from plasma which must take place at the same time occur either in the urine, the gastro-intestinal tract, or the intracellular fluid. The latter is exemplified when there are large potassium losses from the cells. Some cellular K^+ is then exchanged for H^+ and KCl is lost in the urine.

Metabolic acid-base disturbances due to electrolyte changes may occur in a variety of clinical disorders such as:

1 Renal compensation for respiratory acidosis or alkalosis. (See page 65.)
2 Chloride loss via vomitus as in pyloric stenosis. (See page 104.)
3 Chloride loss via urine
   a) Together with potassium when there is an overproduction of aldosterone.
   b) As a result of diuretics such as mercury preparations and thiazides. (See page 107.)
4 Sodium bicarbonate losses in the stools in diarrhoea. (See page 102.)
5 Chloride losses via perspiration in stokers' disease.
6 Iatrogenic disturbances due to excessive (mainly parenteral) administration of sodium or chlorides.

In the treatment of these cases it is more important to substitute the lost electrolytes than to concentrate on the acidosis or alkalosis. Only if the pH changes are excessive is there any cause for the immediate correction of the acid-base disturbance.

In this connection it should be remembered that the body endures a pH decrease better than a pH increase. A practical rule is to correct towards normal a pH which is below 7.15 or above 7.55. If the pH change is due to an electrolyte disturbance, we have to know how much to give. The apparently correct formula for substituting an ion deficiency is:

$$\text{mmol needed} = \text{Body weight (kg)} \times 15\% \times \Delta \text{ ion conc}$$

where 15 is an approximate figure for the extracellular fluid volume and Δ ion conc is the difference between the normal ion concentration and the actual level found, as shown in the following example:

A man weighing 70 Kg has a metabolic alkalosis because of vomiting. His Cl_p^- is 85 mmol/litre. How much chloride does he need for a normalisation of his Base excess?

$$70 \times 15/100 \times (101-85) = 168 \text{ mmol}$$

The calculation is far less exact than it appears because, as repeatedly stated, water changes occur ordinarily together with electrolyte losses and no account is taken of this in the formula. Moreover, the formula may be used only for sodium and chloride. Potassium is located mainly intracellularly and if a man of 70 kg has a K_p^+ of 3.0 mmol/litre, it does not indicate that he needs $70 \times 15/100 \times 1 = 10$ mmol of potassium. He may as well need 100 or 1000 mmol.

Again, the formula may not be used for correction of HCO_3^- as this concentration depends upon the availability of cations. Sodium bicarbonate is often used to correct a sodium deficiency.

The best way to treat acid-base disturbances due to electrolyte losses is to measure the losses and to substitute accordingly. Thus, when possible, save all the fluid losses and check their volume and electrolytes. Often, however, this is not possible and it must suffice to estimate the volume of the losses and to judge the electrolyte concentrations from earlier experience. Table 6, page 102, shows how wide the range may be from case to case. As the electrolyte concentration varies greatly such tables are unreliable and repeated plasma electrolyte determinations are needed to ensure correct treatment.

Table 6

Mean electrolyte concentrations in mmol/litre in different body fluids in adult. ( ) indicates range. Often the variations are so large that no representative mean should be given.

	Na$^+$	K$^+$	Cl$^-$
Vomitus	60	10	
	(10–116)	(1–33)	(78–159)
Loose stools	80	21	48
	(45–125)	(4–48)	(18– 89)
Urine	(40– 90)	(26–60)	(40–120)
	45	5	58
Perspiration	(18– 97)	(3–12)	(18– 97)

The table is compiled from references 5, 9, and 17.

It may be concluded from the above that the electrolyte losses come mainly from the extracellular fluid, but partly from the intracellular. The total fluid losses may usually be gauged from the changes in the body weight of the patient.

A patient who appears to be dehydrated has usually lost some 5 % of his total body water and in extreme cases 10 %. About half of these losses come from the intra- and half from the extracellular fluid compartment. The electrolyte substitution may also be estimated from the normal plasma levels (Fig. 11, page 36) and from the intracellular levels given on page 41.

Clinical examples

Metabolic acidosis

An eight-week-old child weighing four kg was brought to the hospital with a one-day history of fever and watery stools five times. Physical examination revealed no abnormalities except for mild dehydration. He was given fluid orally and appeared to be doing well when suddenly after a few hours he produced nine watery stools within a period of four hours. Blood analyses at that time showed:

Na_p^+	150 mmol/litre
Cl_p^-	128 ,,
K_p^+	5.3 ,,
Total CO_{2p}	8 ,,
Hb	90 g/litre

As the infant had none of the clinical entities which go with increased concentrations of residual anions, the metabolic component may be estimated from $Na_p^+ - Cl_p^-$. (See also page 39.) Buffer base then is $150-128=22$ mmol/litre and, assuming a normal BB_p of 42 mmol/litre, BD_p is $42-22=20$ mmol/litre, Roughly the same information is obtained from the direct analysis of Total CO_2. In this case it would have been better to measure pH than Total CO_2, but let us try to estimate pH. If Total CO_2 8 is connected with Pco_2 40 mm Hg in a Siggaard-Andersen alignment nomogram pH is found to be 6.90. This is a very low value; probably the infant had some degree of respiratory compensation. According to Fig. 19, Pco_2 at BD 20 mmol/litre would have been 16 mm Hg and pH 7.33. It is hardly to be expected that the infant had time to develope a maximal compensation. If instead we try to connect Total CO_2 8 with BD 20, we find a pH of 7.03 and a Pco_2 of 30 mm Hg, and these values appear more likely. Although we cannot know the true values, it is clear that the patient had a severe metabolic acidosis and a low pH. The cause for this was the large amount of HCO_3^- he lost via the stools. Most of the bicarbonate excreted in the faeces is bound to sodium, and only the HCO_3^- molecule affects the acid-base balance. This is the reverse situation from that in renal compensation for respiratory acidosis, when there is a HCO_3^- retention (page 66).

The laboratory findings in this case not only give information about the electrolytes and the acid-base balance, they also clearly indicate fluid disturbances. As mentioned for instance on page 99, the Na_p^+ often gives more information about total water volume than about the amount of sodium. In this case there was a slight increase in the sodium level as a sign of some dehydration, but, in relation to the sodium loss, the water loss is of the same order of magnitude, as Na_p^+ is so close to the normal value. In this case the high chloride concentration gives better information about the total fluid loss from the extracellular fluid compartment than does Na_p^+. As the chloride anion is limited mainly to the extracellular space, an increase in Cl_p^- from 101 to 128 mmol/litre corresponds to a 20 % reduction of the extracellular fluid volume.

Such a calculation is easy and sometimes useful in order to obtain some estimation of the size of the fluid losses.

Assume that the extracellular fluid volume is a litre, its total chloride content is b mmol, that its fluid loss is x percent, and the initial chloride concentration in the plasma was 100 mmol/litre.

We then have $b = 100$ a initially and
$b = 128(a - ax/100)$ after the fluid loss.
then $100 a = 128 (1-x)$ and $x = 20\%$.

Actually hereby the real loss is underestimated because it is assumed that the total chloride is constant. It is no problem to correct the calculations in cases of chloride losses: it suffices to give the final b a factor, say 0.9, depending upon what fluid losses the patient had and the values in Table 6.

Remember also the potassium losses in these cases. Table 6 indicates that the potassium losses are about 1/4 of those of sodium. As soon as the infant is rehydrated and the metabolic acidosis is abolished, K_p^+ will fall rapidly. In the present case a K_p^+ of 2.1 mmol/litre was noted on the first day after admission, although by then all the other electrolyte values were normal.

In summary, a pH value would have been useful, otherwise the sodium and chloride figures suffice for an estimation of the acid-base and fluid balance.

Metabolic alkalosis

A man 58 years of age with a previous history of repeated duodenal ulcers complains of pains in the epigastrium after eating rich Christmas food and vomiting consistently after each meal. After ten days of this he was admitted to hospital, weak, dry, looking very ill and in a somewhat confused condition. The following laboratory information was given:

Na_p^+	141 mmol/litre
Cl_p^-	75 ,,
K_p^+	2.3 ,,
pH	7.55 (H^+ 28 nanomol/litre)
Pco_2	60 mm Hg (8.0 kPa)
BE_{ECF}	26.5 mmol/litre
BE_p	28.0 ,,
Hb	150 g/litre

In this case all the measurements fail to give us appropriate information about the total fluid loss, although clinically it was apparent that the patient was dehydrated. A reliable Hb value obtained before the onset of the present sickness may have been useful. As it was, the "normal" Hb was the result of a low Hb + dehydration.

Outstanding among the electrolytes are the low Cl_p^- and the low K_p^+. As seen from Table 6 the patient has lost chloride, potassium, and sodium by his vomitus. All the chlorides come from the extracellular fluid compartment. The sodium changes are not so large because in order to maintain a normal osmolarity the body adjusts its water loss to the sodium loss, or, alternatively, its sodium loss via the kidney to the water balance of the body. For the latter reasons the sodium deficiency is minimized and the patient needs mainly potassium and chloride.

Cl^- is lost in the vomitus together with the cations Na^+, K^+ and H^+. The amount of H^+ may be measured from titrations, but also from Cl^-—$(K^+ + Na^+)$. The potassium, as always, comes from the intracellular compartment and H^+ from the buffer systems. When the anion Cl^- is lost HCO_3^- is available. The extracellular alkalosis depends therefore upon either a pure H^+ loss via the vomitus or an exchange between the extra- and intracellular compartments. K^+ escapes and H^+ enters the cells.

As this patient has none of the diseases listed on page 39 leading to increased residual anion concentrations, BB and BE may also be calculated from the electrolytes, i.e. $Na_p^+ - Cl_p^-$. We find in this case $141 - 75 = 66$, which then corresponds to Buffer base. Base excess$_p$ is $66 - 42 = 24$ mmol/litre. As seen above, this agrees not too badly with the observed 28 mmol/litre. It could therefore be argued that the acid-base measurements were superfluous, but this was not so as they gave the following important points:

1 pH was 7.55 ($H^+ = 28$ nanomol/litre)
2 Pco$_2$ was 60 mm Hg (8.0 kPa)

ad 1 pH is so high that a symptomatic correction is needed. Give, for instance, 50 mmol chloride in the form of NH$_4$Cl. As stated above a corresponding amount of H^+, ie. 50 mmol will be liberated in the liver. This ammonium chloride infusion should increase Cl_p^- about 5 mmol/litre and reduce BE correspondingly. If Pco$_2$ remains unchanged, pH would fall from 7.55 to 7.50 as may be read from the alignment nomogram.

These calculations may need some explaining. This patient weighs 70 kg and should have an extracellular fluid volume of 17 % of body weight. Because of the loss of fluid we estimate his extracellular fluid compartment to 15 % of his weight, i.e. $15/100 \times 70 = 10$ litres. The 50 mmol Cl^- the patient was given will stay within the fluid volume of 10 litres; consequently the increase is 5 mmol/litre.

After the decrease in pH the patient will improve enough to give the doctor time to correct the fluid and electrolyte balance without having to bother with the acid-base status. It would now be suitable to give this patient either Saline or Ringer solution because of the relatively high chloride concentration in both. Finally, potassium is needed as well. As soon as pH is reduced, K_p^+ usually increases, but this is no proof that supplementary potassium is not needed.

ad 2 As seen from Fig. 23, page 74, P_{CO_2} could be expected to rise to 60 mm Hg as a compensation for a metabolic alkalosis of 25 mmol/litre. The primary disease is pyloric stenosis with vomiting, the classical cause of metabolic alkalosis, and pH is elevated. The high P_{CO_2} cannot be other than a compensation. Without this, pH would have been 7.70 instead.

When such a sick patient is admitted, alternative treatment must be discussed and an operation should be considered. An anaesthesiologist can be consulted. It is not enough to tell the anaesthesiologist that we have a severely sick patient on whom we may have to operate, but who has a P_{CO_2} of 60 mm Hg. There are, however, cases on record when this has been done, artificial ventilation has been instituted, and dangerously high pH levels have been reached. As always, the complete clinical picture must be taken into account together with the laboratory results; it does not suffice to concentrate on only one of the parameters of the acid-base balance.

Table 6, page 102 gives a mean value of 10 mmol K^+ per litre of vomitus. If the patient thus lost two litres daily for ten days he has lost a total of 200 mmol K^+.

The total needs of the patient then are

1) the 200 mmol K^+ lost
2) his daily K^+ intake
3) his extra potassium losses, if any.

If the patient is given KCl, his potassium deficiency should disappear and Base excess decrease by about 20 mmol/l, using the same type of calculation used earlier for this patient.

The daily potassium intake is about 40 mmol. The extra losses were estimated at 20 mmol/24 hours; thus the maintainance dosage should be 60 mmol/24 hours. The 200 mmol deficit cannot then be given in only one day; given over a period of 4 days the dosage would be 110 mmol/day.

CHAPTER 17

Diuretics

The influence of diuretics on the electrolyte and acid-base balance

In prescribing diuretics the primary aim is to eliminate an excess of fluid, but this will succeed only if at the same time a corresponding amount of electrolytes is lost. Every mmol of sodium holds 7 ml of water so that every mmol sodium lost also means a water loss of 7 ml. The diuretics increase the sodium elimination and as a consequence there will be plasma-electrolyte and acid-base changes.

Most of the diuretics cause metabolic alkalosis, a few cause acidosis.

1 Acidifying diuretics
   Acetazolamide (Diamox®) and Triamterene
2 Alkalizing diuretics
   Mercurials, thiazides and other saluretics

ad 1 Acetazolamide reduces the reabsorption of HCO_3^- in the renal tubuli. HCO_3^- is lost together with the cation Na^+ in the urine. When HCO_3^- is lost, H^+ is retained. This is coupled to the remaining HCO_3^- in the formation of H_2O and CO_2 and the latter is eliminated by ventilation, whereby the plasma HCO_3^- concentration is reduced. Usually Na_p^+ falls because of the relatively larger loss of sodium than that of water, whereas Cl_p^- remains stable. $(Na_p^+ - Cl_p^-)$ is reduced, another sign of metabolic acidosis.

Besides sodium and bicarbonate, the urine after acetazolamide treatment contains more potassium than usual and in spite of falling pH these patients have low K_p^+, i.e. contrary to the rule already cited several times.

During Triamterene administration the situation is slightly different. These patients also lose sodium and bicarbonate, but their metabolic acidosis is never as pronounced as after acetazolamide. Most important, instead of potassium losses these patients have potassium retention to such a degree that uncautious use of the drug may lead to severe hyperkalaemia.

ad 2 Sometimes all the drugs in this group are called saluretics, as their main function is to decrease the sodium reabsorption in the renal tubuli. At the same time there must of course be an elimination of an equal amount of anions, mainly chloride.

An analysis of the major electrolytes in the urine after the administration of a thiazide gave the following representative results:

Na^+ 330 mmol/24 hours
K^+ 30 „
Cl^- 360 „

Thus the relation Na^+/Cl^- in the urine was 0.9, while in plasma it was 1.4 (143/101). In relation to plasma, extra chloride was therefore lost; this chloride in the plasma was replaced by HCO_3^-. After about one week of treatment with a thiazide, Base excess is usually somewhere between 7 and 13 mmol/litre. If the drug treatment is continued, there is usually a gradual normalization of Base excess, but it remains at 5 or 6 mmol/litre (120).

It cannot be repeated too often that the most common cause of metabolic alkalosis today is treatment with diuretic drugs. Whether they are primarily prescribed as diuretics or as hypotensive agents, the same metabolic alkalosis occurs. I have frequently been asked to comment upon "unexplained" metabolic alkalosis and my first guess, i.e. diuretics, has usually been right.

The need for supplementary potassium during therapy with diuretics

Many insist that all patients taking diuretics should have supplementary potassium. The statement would be improved if reworded as: all such patients need potassium *chloride,* as the chloride losses are so important (118). If in these cases potassium is given without chloride no effect should be expected. But in any case the effect is none too good on K_p^+, as seen in Fig. 29, where the crosses mark cases given 3 g of KCl daily and the dots mark cases receiving only the diuretic agent. It will be seen that both groups fall along the same regression line. The main reason for the low plasma potassium level is the metabolic alkalosis; if this is abolished K_p^+ usually becomes normal. Give the patients, for instance, 3 g of NH_4Cl (in coated tablets and with ample fluid, to prevent vomiting).

Many patients taking thiazides for a long time do not need any supplementary KCl, but, on the other hand, the number of cases seen with severe hypokalaemia is increasing. Usually the hypokalaemic patients are found to have heart or liver insufficiency and therefore have marked potassium losses in any case. KCl should be the rule here.

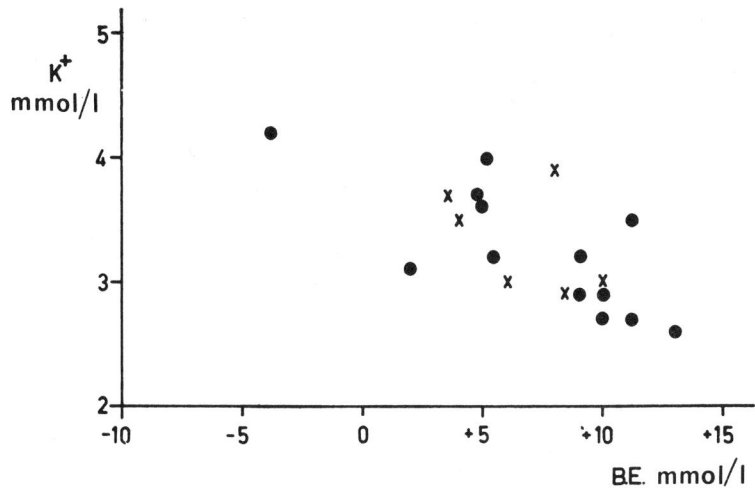

Fig. 29. The relation between plasma potassium and Base excess in patients given thiazides. x stands for patients given supplementary 3 g KCl/24 hours.

In the great majority of cases I prefer to give no extra KCl. The patients usually have enough tablets to take already, not forgetting the cost of the superfluous drug. It is better to instruct the patient, and preferably his relatives as well, that if he becomes tired and/or constipated he should preferably report at once for a K_p^+ analysis, or begin taking KCl tablets.

CHAPTER 18

Salicylate intoxication

Salicylate poisoning is one of the clinical situations in which the residual anion concentration is increased. Initially the patients have pure respiratory alkalosis and gradually a metabolic acidosis ensues; therefore this entity gives a suitable summary of demonstrations of acid-base disturbances.

Primarily there is a hyperventilation initiated by the cerebral action of the drug. The alveolar elimination of CO_2 then becomes larger per time unit than the metabolically produced CO_2, the total body content of CO_2 diminishes, and the physically dissolved CO_2 concentration is reduced i.e. Pco_2 falls.

> Example: A patient has initially normal values
> $pH = 7.40$ ($H^+ = 40$ nanomol/litre)
> $Pco_2 = 40$ mm Hg (5.3 kPa)
> Base excess $= 0$ mmol/litre

When patients begin to hyperventilate, BE_{ECF} remains unchanged but total CO_2 in plasma is reduced, say from the normal level of 25 to 23 mmol/litre. When total CO_2 23 is connected to BE_{ECF} 0, pH is found to be 7.52 and Pco_2, 28 mm Hg. This, as will be seen, is just another way of using the versatile alignment nomogram.

The patients not only hyperventilate; they also have hypermetabolism, i.e. both their oxygen consumption and their CO_2 production and heat production increase. In spite of the increased amounts of CO_2 to be eliminated Pco_2 is low. Both the hyperventilation per se and the increased body temperature increase the water losses and the patients need relatively large amounts of fluids. Often they vomit and cannot drink sufficiently. Remember in these cases always to give more fluid than is apparently needed in relation to the weight of the patient.

After about 24 hours the acid-base status changes and gradually there appears more and more of a metabolic acidosis, although it seems as yet unclear what chemical substance accounts for the increase in residual anions. To some extent there is an increase in salicylate and to some extent, an increase in lactate (see also page 39).

The following data were obtained from a patient who had suffered from salicylate poisoning for 48 hours:

Na_p^+	142 mmol/litre
K_p^+	6.4 ,,
Cl_p^-	96 ,,
BD_{ECF}	21.5 ,,
BD_p	20.0 ,,
Pco_2	32 mm Hg (4.2 kPa)
pH	7.00 (100 nanomol/litre)

The reader should plot these figures in a Gamble diagram. Assuming normal values for calcium and magnesium, total cation concentration is 154 meq/litre. As the sodium and chloride concentrations are more or less normal, the metabolic acidosis can be explained only by an increase in the residual anions. Buffer base is 42—20=22 mmol/litre. The residual anions then are 154— —(96+22) = 36 meq/litre; normally they should be about 10. The low pH of the patient must be corrected and after 1 litre of the 167 mmol $NaHCO_3$ solution Base deficit is reduced by about 10 mmol/litre.

Just as when treating renal insufficiency with sodium bicarbonate, the basic metabolic abnormality remains unchanged and there is only a slow disappearance of the primary metabolic acidosis as the patient improves. Therefore there is no residue of iatrogenic metabolic alkalosis of the sodium bicarbonate therapy in salicylate poisoning.

As stated above, the patient begins with hyperventilation, low Pco_2, and then, of course, high pH, and later develops a metabolic acidosis. At one stage, then, pH is normal because the hyperventilation and the metabolic acidosis balances pH at the normal level of 7.40. Take care not to misinterpret the situation and believe that the patient is cured because of his normal pH. On the contrary, this is when the dangerous phase of the disease begins.

References

General reference

1. *Bunton, G. L.:* Fluid Balance without Tears. Lloyd-Luke, London (1965).
2. *Creese, R., Neil, M. W., Ledingham, J. M.* and *Vere, D. W.:* The Terminology of Acid-Base Regulation. Lancet I:419 (1962).
3. Current Concepts of Acid-Base Measurement. Annals of the New York Academy of Science 133 (1966).
4. *Davenport, H.:* The ABC of Acid-Base Chemistry. University of Chicago Press, Chicago (1955).
5. *Elkinton, J. R.* and *Danowski, T. S.: The Body Fluids:* Basic Physiology and Practical Therapeutics. Williams & Wilkins Co., Baltimore (1955).
6. *Gamble, J. L.:* Chemical Anatomy, Physiology and Pathology of Extracellular Fluid. Harvard University Press. Cambridge, Mass. (1950).
7. *Henderson, L. J.:* Blood. Yale University Press, New Haven (1928).
8. *Kildeberg, P.:* Clinical Acid-Base Physiology. Studies in Neonates, Infants and Young Children. Williams and Wilkins Co., Baltimore (1968).
9. *Maxwell, M. H.* and *Kleeman, C. R.:* Clinical Disorders of Fluid and Electrolyte Metabolism. McGraw-Hill, New York (1962).
10. *Muntwylen, E.:* Water and Electrolyte Metabolism and Acid-Base Balance. C. V. Mosby, USA (1971).
11. *Peters, J. P.* and *van Slyke, D. D.:* Quantitative Clinical Chemistry I. The Williams & Wilkins Company, Baltimore (1931).
12. *Rooth, G.:* Acid-Base and Electrolyte Balance. Vol 1 Introduction. Studentlitteratur, Lund, Sweden 6th Ed (1972).
13. Idem Vol 2 Clinical. Studentlitteratur, Lund, Sweden (1970).
14. *Saling, E.:* Das Kind im Bereich der Geburtshilfe. Georg Thieme, Stuttgart (1966).
15. *Siggaard-Andersen, O.:* The Acid-Base Status of the Blood. Copenhagen (1963).
16. *Thorén, L.:* Vätskebalans. Almqvist & Wiksells, Uppsala (1960).
17. *Weisberg, J. F.:* Electrolyte and Acid-Base Balance. Williams & Wilkins, Baltimore (1962).
18. *Winters, R. W., Engel, K.* and *Dell, R. B.:* Acid Base Physiology in Medicine. The London Comp. of Cleveland and Radiometer A/S Copenhagen (1967).
19. *Woolmer, R. F.:* Ed. pH and Blood Gas Measurement. J. A. Churchill, London (1959).

References, Chapters 1—5

20 *Astrup, P.:* A simple electrometric technique for the Determination of Carbon Dioxide Tension. Scand. J. clin. Lab. Invest. 8:33 (1956).

21 *Astrup, P., Jørgensen, K., Siggaard-Andersen, O.* and *Engel, K.:* The Acid-Base Metabolism. A New Approach. Lancet I:1035 (1960).

22 *Brönsted, J. N.:* The Conception of Acids and Bases. Rec. trev. chim. 42:718 (1923).

23 *Dormandy, T. L.:* Body pH. Lancet I:755 (1966).

24 *Edelman, I. S.* and *Leibman, J.:* Anatomy of Body Water and Electrolytes. Am. J. Med. 27:256 (1959).

25 Editoral. Hydrogen Ions and Buffer Base. Am. J. Med. 25:1 (1958).

26 *Eliasson, R.:* Kroppsvätskornas Osmolalitet – en Faktor att Räkna med. Läkartid. 69:2321 (1972).

27 *Harrington, J. T.* and *Lemann, J. J.:* The Metabolic Production and Disposal of Acid and Alkali. Med. Clin. N. Amer. 54:1543 (1970).

28 *Kaplan, S. A.:* Fluid and Electrolyte Therapy: Maintenance, Abnormal Status, Methods of Administration. Pediat. Clin. N. Amer. 16:581 (1969).

29 *Mellengaard, K.* and *Astrup, P.:* The Quantitative Determination of Surplus Amounts of Acid or Base in the Human Body. Scand. J. clin. Lab. Invest. 12:187 (1960).

30 *Lawin, P.:* Störungen des Säure-Basen-Haushaltes: Differentialdiagnose und Therapie. Dtsch. med Wschr. 93:1664 (1968).

31 *Müller-Plathe, O.:* Die Behandlung der metabolischen Azidose. Dtsch. med. Wschr. 93:1661 (1968).

32 *Owen, J. A., Dudley, H. A. F.* and *Masterton, J. P.:* Acid-Base Status Assessed from Measurements of Hydrogen Ion Concentration and P_{CO_2}. Lancet II:660 (1965).

33 *Rahn, H.:* Gas Transport from the External Environment to the Cell. in: Development of the Lung, Ciba Foundation Symposium. J. A. Churchill, London (1967).

34 *Redstone, D.* and *Beard, R. W.:* Correction for Oxygen Desaturation as a Source of Error in the Determination of Blood P_{CO_2} by the Astrup Interpolation Method. Clin. Chim. Acta, 27:317 (1970).

35 *Severinghaus, J. W.:* Blood Gas Concentrations. in: Handbook of Physiology, section 3, Respiration Vol. II, chapter 61, p. 1477. American Physiological Society, Washington (1965).

36 *Severinghaus, J. W.* and *Bradley, A. F.:* Electrodes for Blood P_{O_2} and P_{CO_2} Determination. J. appl. Physiol. 13:515 (1958).

37 *Severinghaus, J. W.* and *Bradley, A. F.:* Blood Gas Electrodes or What the Instructions Didn't Say. Radiometer. Copenhagen (1971).

38 *Siggaard-Andersen, O.:* Acid-Base Disturbances. Lancet I:1104 (1964).

34 *Siggard-Andersen, O.:* Acute Experimental Acid-Base Disturbances in Dogs. Scand. J. clin. Lab. Invest. 14:suppl. 66 (1962).

40 *van Slyke, D. S., Hastings, A. B., Hiller A.* and *Sendroy, J., Jr.:* The Amount of Alkali Bound by Serum Albumin and Globalin. J. Biol. Chem. 79:769 (1928).

41 *Staff, P. H.* och *Nilsson, S.:* Vaeske- og Sukkertilförsel under långvari intens fysisk aktivitet. T. norske Laegeforen. 91. 1235 (1971).

42 *Weil, W. B.:* A Unified Guide to Parenteral Fluid Therapy. I Maintenance Requirements and Repair of Dehydration. J. Pediat. 75:1 (1969).

43 *Whithead, T. P.:* Acid-Base Status, pH, and Pco_2. Lancet II:1015 (1965).

44 *Winters, R. W.:* Studies of Acid-Base Disturbances. Pediatrics 39:700 (1967).

45 *Wulf, H., Künzel, W.* and *Lehmann, V.:* Vergleichende Untersuchungen der aktuellen Blutgase und des Säure-Basen-Status im fetalen und maternen Kapillarblut während der Geburt. Z. Geburtsh. Gynaek. 167:113 (1967).

References, Chapters 6—9

46 *Bleich, H. L.:* Computer Evaluation of Acid-Base Disorders. J. clin. Invest. 48:1689 (1969).

47 *Briggs, A. P.:* Erythrocyte Mechanisms in the Transport of Carbon Dioxide: An Optional Scheme. Metabolism 17:582 (1968).

48 *Dell, R. B., Lee, C. F.* and *Winters, R. W.:* Effects of Expansion of the Volume of ECF on the in Vivo CO_2 "Titration Curve". Fed. Proc. 27:509 (1968).

49 *Bonham, T. J.* and *Sammons, H. G.:* The Effect of Po_2 on the Determination of Pco_2 by the interpolation technique. Clin. Chim. Acta 29:507 (1970).

50 *Bracket, N. C., Cohen, J. J.* and *Schwartz, W. B.:* Carbon Dioxide Titration Curve of Normal Man. New Engl. J. Med. 272:6 (1965).

51 *Englesson, S., Grevsten, S.* and *Olin, A.:* Some Numerical Methods of Estimating Acid-Base Variables in Normal Human Blood with a Haemoglobin Concentration of 5 g/100 cm^3. Scand. J. clin. Lab. Invest. 32:289 (1973).

52 *Graham, J. A., Lamb, J. F.* and *Linton, A. L.:* Measurement of Body Water and Intracellular Electrolytes by Means of Muscle Biopsy. Lancet II:1172 (1967).

53 *Huckabee, W. E.:* Relationships of Pyruvate and Lactate during Anaerobic Metabolism. I. Effects of Infusion of Pyruvate or Glucose and of Hyperventilation. J. clin. Invest. 37:244 (1958).

54 *Langlands, J. H. M.* and *Wallace, W. F. M.:* Small Blood Samples from Ear-Lobe puncture. A Substitute for Arterial Puncture. Lancet II:315 (1965).

57 *Lindenbaum, J., Akbar, R., Gordon, R., Greenough III, W. B., Hirschorn, N.* and *Islam, M. R.:* Cholera in Children. Lancet I:1066 (1966).

56 *Lotspeich, W. D.:* Metabolic Aspects of Acid-Base Change: Science 155:1066 (1967).

57 *MacIntyre, J., Norman, J. N.* and *Smith, G.:* Use of Capillary Blood in Measurement of Arterial pO_2. Brit. med. J. II:640 (1968).

58 *Rooth, G.* och *Jacobson, L.:* The Value and Validity of Base Excess$_{ECF}$ in Perinatal Acid-Base Studies. Scand. J. clin. Lab. Invest. 28:283 (1971).

59 *Rooth, G.* and *Thalme, B.:* The Validity of Buffer Base and Base Excess in Acid-Base Studies. Amer. J. Obstet. Gynec. 108:282 (1970).

60 *Schwartz, W. B.* and *Relman, A. S.:* A Critique of the Parameters used in the Evaluation of Acid-Base Disorders. New. Engl. J. Med. 268:1382 (1963).

61 *Siggaard-Andersen, O.* and *Engel, K.:* A New Acid-Base Nomogram. An Improved Method for the Calculation of the Relevant Blood Acid-Base Data. Scand. J. clin. Lab. Invest. 12:177 (1960).

62 *Siggaard-Andersen, O.:* Blood Acid-Base Alignment Nomogram. Scand. J. clin. Lab. Invest. 15:211 (1963).

63 *Siggaard-Andersen, O.:* Therapeutic Aspects of Acid-Base Disorders. In: Modern Trends in Anaesthesia 3:99. Evans and Gray Ed., Butterworths, London (1966).

64 *Singer, R. B.* and *Hastings, A. B.:* An Improved Clinical Method for the Estimation of Disturbances of the Acid-Base Balance of Human Blood. Medicine 27:2223 (1948).

65 *Stamm, S. J.:* Reliability of Capillary Blood for the Measurement of Po$_2$ and O$_2$ Saturation. Dis. Chest 52:191 (1967).

66 *Stinebauch, B., Miller, R. B.* and *Relman, A. S.:* The Influence of Non-reabsorbable Anions on Acid Excretion. Clin. Sci. 36:53 (1969).

67 *Sutton, R. N., Wilson, R. F.* and *Walt, A. J.:* Differences in Acid-Base Levels and Oxygen-Saturation between Central Venous and Arterial Blood. Lancet II: 749 (1967).

References, Chapters 10—11

68 *Arbus, G. C., Hebert, L. A., Levesque, P. R., Etsten, B. E.* and *Schwartz, W. B.:* Characterization and Clinical Application of the "Significance band" for Acute Respiratory Alkalosis. New Engl. J. Med. 280:117 (1969).

69 *Boutourline-Young, H.* and *Boutourline-Young, E.:* Alveolar Carbon Dioxide Levels in Pregnant, Parturient and Lactating Subjects. J. Obstet. Gynaec. Brit. Emp. 63:509 (1956).

70 *Brackett, N. C., Wingo, C. F., Muren, O.* and *Solano, J.:* Acid-Base Response to Chronic Hypercapnia in Man. New Engl. J. Med. 280:124 (1969).

71 *Böning, D.* and *Heinrich, K. W.:* Veränderungen der CO$_2$-Bindungskurve des Blutes bei akuter respiratorischer Acidose und ihre Ursachen. II. Untersuchungen am Menschen. Pflügers Arch. ges. Physiol. 303:162 (1968).

72 *Dawnes, J. J., Wood, D. W., Striker, T. W.* and *Pittman, J. C.:* Arterial Blood gas and Acid-Base Disorders in Infants and Children with Status Asthmaticus. Pediatrics 42:238 (1968).

73 *Dulfano, M. J.* and *Ishikawa, S.:* Quantitative Acid-Base Relationships in Chronic Pulmonary Patients during the Stable State. Amer. Rev. resp. Dis. 93:251 (1966).

74 *Engel, K., Dell, R. B., Rahill, W. J.* and *Winters, R. W.:* Quantitative Displace-

ment of Acid-Base Equilibrium in Chronic Respiratory Acidosis. J. appl. Physiol. 24:288 (1968).

75 *Engel, K., Kildeberg, P.* and *Winters, R. W.:* Quantitative Displacement of Blood Acid-Base in acute Hypocapnia. Scand. J. clin. Lab. Invest. 23:5 (1969).

76 *Forwand, S. A., Landowne, M., Follansbee, J. N.* and *Hansen, J. E.:* Effect of Acetazolamide on Acute Mountain Sickness. New Engl. J. Med. 279:839 (1968).

77 *Miller, A., Teirstein, A. S., Duberstein, J., Chusid, E. L., Bader, M. E.* and *Bader, R. A.:* Use of Oxygen Inhalation in Evaluation of Respiratory Acidosis in Patients with Apparent Metabolic Alkalosis. Amer. J. Med. 45:513 (1968).

78 *McNical, M. W.* and *Campbell, E. J. M.:* Severity of Respiratory Failure. Lancet I:337 (1965).

79 *Refsum, H. E.:* Acid-Base Status in Patients with Chronic Hypercapnia and Hypoxaemia. Clin. Sci. 27:407 (1964).

80 *Refsum, H. E.:* Acid-Base Disturbances in Chronic Pulmonary Disease. In: Current Concepts of Acid-Base Measurement, ed. E. M. Weyer. Ann. N. Y. Acad. Sci. 133: (1966).

81 *Rossier, P. H.* and *Hotz, M.:* Respiratorische Funktion und Säurebasengleichgewicht in der Schwangerschaft. Schweiz. med. Wschr. 83:897 (1953).

82 *Schwartz, W. B., Bracket, N. C.* and *Cohen, J. J.:* The Response of Extracellular Hydrogen Ion Composition to graded Degress of chronic Hypercapnia: the Physiologic Limits of the Defence of pH. J. Clin. Invest. 44:291 (1965).

83 *Sluiter, H. J., Koolhaas, B., van der Lende, R., Tammeling, G. J., Blokzijl, E., van Dijl, W.* and *Orie, N. G. M.:* Conservative and aggressive Treatment of acute severe Respiratory Insufficiency in Patients with Chronic Non-Specific Lung Disease (C. N. S. L. D.). Med. thorac. 21:335 (1964).

84 *Sluiter, H. J., Tammeling, G. J.* and *Orie, N. G. M.:* The Treatment of acute Respiratory Failure in Patients with Chronic Aspecific Respiratory Affections (CARA). Sel. Pap. Roy. Netherl. Tuber. Ass., The Hague VII:5 (1963).

85 *Weiss, E. B.* and *Dulfano, M. J.:* Quantitative Acid-Base Dynamics in chronic Pulmonary Disease. Ann. intern. Med. 69:263 (1968).

References, Chapters 12—14

86 *Albert, M. S., Dell, R. B.* and *Winters, R. W.:* Quantitative Displacement of Acid-Base Equilibrium in Metabolic Acidosis. Ann. intern. Med. 66:312 (1967).

87 *Alberti, K. G. M. M., Hockaday, T. D. R.* and *Turner, R. C.:* Small Doses of Intramuscular Insulin in the Treatment of Diabetic "Coma". Lancet II:515 (1973).

88 *Johnson, R. D., Conn, J. W., Dykman, C. J., Pek, S.* and *Starr, J. I.:* Mechanisms and Management of Hyperosmolar Coma without Ketoacidosis in the Diabetic. Diabetes 18:11 (1969).

89 *Krebs, H. A.:* Bovine Ketosis. Vet. Record 78:187 (1966).

90 MacCurdy, D. K.: Hyperosmolar Hyperglycemic Nonketotic Diabetic Coma. Med. Clin. N. Amer. 54:683 (1970).

91 Monti, M and Rooth, G.: Respiratory Compensation to Metabolic Acid-Base Disturbances. Scand. J. clin. Lab. Invest. 26:381 (1970).

92 Huckabee, W. E.: Abnormal resting Blood Lactate. I. The Significance of Hyperlactatemia in hospitalized Patients (see Reference nr 93).

93 Huckabee, W. E.: Abnormal resting Blood Lactate. II. Lactic Acidosis. Amer. J. Med. 30:833, (1961).

94 Oakes, D. D., Schreiman, P. H., Hoffman, R. S. and Arky, R. A.: Hyperglycemic, Non-Ketotic Coma in the Patient with Burns: Factors in Pathogenesis. Metabolism. 18:103 (1969).

95 Pullen, H., Doig, A. and Lambie, A. T.: Intensive Intravenous Potassium Replacement Therapy Lancet II:809 (1967).

96 Robin, E.: Attention Called to Danger of an Alkaline Overshoot with 'Life-Saving' Bicarbonate for Lactic Acidosis. Diabetes Outlook 8:1 (1973).

97 Rooth, G.: The time Factor in Foetal Distress. J. Perinat. Med. 1:1 (1973).

98 Rooth, G. and Carlström, S.: Diurnal Variations in Blood Glucose, 3-Hydroxybutyrate, Acetoacetate, Plasma Free Fatty Acids and Glycerol in Diabetics. Acta med. scand. 191:559 (1972).

99 Seftel, H. C., Goldin, A. R. and Rubenstein, A. H.: Hyperosmolar Non-Ketotic Coma. Lancet II:1042 (1967).

100 Soler, N. G., Bennet, M. A., Dixon, K., FitzGerald, M. G. and Malins, J. M.: Potassium Balance during Treatment of Diabetic Ketoacidosis. Lancet II:665 (1972).

101 Vinik, A., Seftel, H. and Joffe, B. I.: Metabolic Findings in Hyperosmolar, Non-Ketotic Diabetic Stupor. Lancet II:797 (1970).

References, Chapter 15

102 Cove-Smith, J. R. and Knapp, M. S.: Sodium Handling in Analgesic Nephropathy. Lancet II:70 (1973).

103 Donadio, J. V.: Conservative Management of Chronic Renal Failure. M. Clin. North America 50, 1175 (1966).

104 Graham, J. A., Lawson, D. H. and Linton, A. L.: Muscle Biopsy Water and Electrolyte Contents in Chronic Renal Failure. Clin. Sci. 38:583 (1970).

105 Kneptschield, J. H., Stone, W. J., Cirksena, W. J.: The Diagnosis and Management of Acute Renal Insufficiency. The Capsule 2, 25 (1971).

106 Papper, S.: A Comprehensive Discussion of Pathophysiology, Clinical Manifestations, and Therapy of Acute and Chronic Kidney Dysfunction. M. Clin. North America 55:335 (1971).

107 Rigolosi, R. S., Frascino, J. A.: Acute Renal Failure and Drug Intoxication. M. Clin. North America 55:1249 (1971).

108 *Rooth, G.* and *Bandmann, U.:* Renal Response to Acid Load after Phenformin. Brit. med. J. 4:256 (1973).

109 *Silverberg, D. S.:* The Use of Mannitol in Oliguric Renal Failure. M. Clin. North America 50:1159 (1966).

110 *Wrong, O.:* Management of the Acute Uraemic Emergency. Br. med. Bull. 27:97 (1971).

111 *Wrong, O., Davies, H. E. F.:* The Excretion of Acid in Renal Disease. Quart. J. Med. 28:259 (1959).

References, Chapters 16—18

112 *Clark, R. G.* and *Norman, J. N.:* Metabolic Alkalosis in Polyric Stenosis. Lancet I:1244 (1964).

113 *Dollery, T., Parry, E. H.* and *Young, D. S.:* Diuretic and Hypotensive Properties of Ethacrynic Acid: a Comparison with Hydrochlorothiazide. Lancet I:947 (1964).

114 *Down, P. F., Polak, A., Rao, R.* and *Mead, J. A.:* Fate of Potassium Supplements in Six Outpatients receiving Longterm Diuretics for Oedematous Disease. Lancet II:721 (1972).

115 Editorial: Diarrhoea and Acid-Base Disturbances. Lancet I:1305 (1966).

116 *Ginsberg, J. D., Saad, A.* and *Gabuzda, G. J.:* Metabolic Studies with the Diuretic Triamterene in Patients with Cirrhosis and Ascites. New Engl. J. Med. 271:1229 (1964).

117 *de Graeff, J. Struyvenberg, A.* and *Lameijer, L. D. F.:* The Role of Chloride in Hypokalemic Alkalosis. Am. J. Med. 37:778 (1964).

118 *Kassirer, J .P.* and *Schwartz, W. B.:* Correction of Metabolic Alkalosis in Man without Repair of Potassium Deficiency. Amer. J. Med. 40:19 (1966).

119 *Pierce, N. F., Banwell, J. G., Mitra, R. C., Bradley Sack, R., Brigham, K. L., Fedson, D. S., Mondal, A.* and *Manji, P. M.:* Oral Replacement of Water and Electrolyte Losses in Cholera. Indian J. Med. Res. 57:848 (1969).

120 *Rooth, G.* and *Fürst, C.:* The Relation between Hypopotassaemia and Alkalosis during Administration of Polythiazide and Chlorthalidone. Acta med. scand. 176:51 (1964).

121 *Sherlock, S., Senewirathne, B., Scott, A.* and *Walker, J. G.:* Complications of Diuretic Therapy in Hepatic Cirrhosis. Lancet I:1446 (1966).

122 *Watten, R. H., Gutman, R. A.* and *Fresh, J. W.:* Comparison of Acetate, Lactate and Bicarbonate in Treating the Acidosis of Cholera. Lancet II:512 (1969).